Materials Horizons: From Nature to Nanomaterials

Series Editor

Materials are an indispensable part of human civilization since the inception of life on earth. With the passage of time, innumerable new materials have been explored as well as developed and the search for new innovative materials continues briskly. Keeping in mind the immense perspectives of various classes of materials, this series aims at providing a comprehensive collection of works across the breadth of materials research at cutting-edge interface of materials science with physics, chemistry, biology and engineering.

This series covers a galaxy of materials ranging from natural materials to nanomaterials. Some of the topics include but not limited to: biological materials, biomimetic materials, ceramics, composites, coatings, functional materials, glasses, inorganic materials, inorganic-organic hybrids, metals, membranes, magnetic materials, manufacturing of materials, nanomaterials, organic materials and pigments to name a few. The series provides most timely and comprehensive information on advanced synthesis, processing, characterization, manufacturing and applications in a broad range of interdisciplinary fields in science, engineering and technology.

This series accepts both authored and edited works, including textbooks, monographs, reference works, and professional books. The books in this series will provide a deep insight into the state-of-art of Materials Horizons and serve students, academic, government and industrial scientists involved in all aspects of materials research.

Review Process

The proposal for each volume is reviewed by the following:

1. Responsible (in-house) editor
2. One external subject expert
3. One of the editorial board members.

The chapters in each volume are individually reviewed single blind by expert reviewers and the volume editor.

Kritesh Kumar Gupta · Sudip Dey ·
Tanmoy Mukhopadhyay

Machine Learning in Nanoscale Materials Design

From Basics to Algorithm Implementation

Springer

Kritesh Kumar Gupta
Amrita School of Artificial Intelligence
Amrita Vishwa Vidyapeetham
Coimbatore, Tamil Nadu, India

Sudip Dey
Department of Mechanical Engineering
National Institute of Technology Silchar
Silchar, Assam, India

Tanmoy Mukhopadhyay
Department of Aeronautics
and Astronautics
University of Southampton
Southampton, UK

ISSN 2524-5384 ISSN 2524-5392 (electronic)
Materials Horizons: From Nature to Nanomaterials
ISBN 978-981-95-2659-8 ISBN 978-981-95-2660-4 (eBook)
https://doi.org/10.1007/978-981-95-2660-4

This Springer imprint is published by the registered company Springer Nature Singapore Pte Ltd.
The registered company address is: 152 Beach Road, #21-01/04 Gateway East, Singapore 189721, Singapore

Preface

Who Should Read It and Why? This book, *Machine Learning in Nanoscale Materials Design: From Basics to Algorithm Implementation*, is ideal for researchers, post-graduate students, and professionals in materials science, chemistry, physics, and computational science who are interested in leveraging the power of machine learning to revolutionize the way nanomaterials are understood and designed. It offers a comprehensive guide to integrating machine learning techniques with molecular simulations, making it especially valuable for those looking to overcome the limitations of traditional modelling methods in handling complex systems at the nanoscale. Whether you're new to the field or seeking to deepen your expertise, this text provides both the theoretical foundations and practical insights necessary to harness the predictive power and efficiency of ML-driven simulations in nanomaterials research.

The rapid advancement of nanotechnology has ushered in an era where the design and manipulation of materials at the atomic and molecular scales are not only possible but critical to innovation across fields such as electronics, energy, biomedicine, and environmental science. Traditional molecular simulations have played a foundational role in this progress, yet their inherent limitations—particularly in terms of scalability and computational cost—present significant barriers in the study of increasingly complex nanoscale systems. In recent years, the emergence of machine learning (ML) as a powerful tool for data-driven modelling and prediction has opened new avenues for overcoming these challenges. This book aims to provide a comprehensive and accessible introduction to the integration of machine learning with molecular simulation techniques, specifically tailored for nanomaterials research. It brings together foundational concepts, state-of-the-art algorithms (including codes that may be readily implemented), and practical case studies to guide readers through this rapidly evolving interdisciplinary landscape. Our goal is to equip scientists, engineers, research scholars, and students with the knowledge and tools necessary to accelerate discovery, optimize material properties, and deepen our understanding of nanoscale phenomena through the synergy of computational modelling and intelligent data analysis.

Existing Works and the Need for This Book: Over the past decade, significant strides have been made in both molecular simulations and machine learning, with each field contributing immensely to scientific progress. Existing literature has explored the individual strengths of these domains—molecular dynamics and quantum simulations offering deep physical insights, while machine learning brings unprecedented speed and predictive accuracy to complex data-driven tasks. A growing number of studies have demonstrated the promise of integrating these techniques, particularly in the context of nanomaterials, where understanding behaviour at the atomic scale is essential for innovation. However, current resources often address this integration in fragmented ways—either focusing on ML methods without adequate grounding in molecular simulation or treating nanomaterials in a limited scope. This book seeks to bridge that gap by providing a unified, coherent treatment of machine learning-based molecular simulation specifically tailored to the challenges and opportunities in nanomaterials research. By offering a structured overview of methods, tools, and applications—along with real-world case studies—this book fills a critical need for a comprehensive, interdisciplinary reference that empowers both newcomers and seasoned researchers to fully harness the potential of this transformative approach.

What Will You Find in This Book? This book covers the essential fundamentals, applications, and important references related to different machine learning approaches specifically applicable to the aspect of Molecular Simulation of Nanomaterials. This book is a structured and in-depth exploration of how machine learning can be effectively integrated with molecular simulation techniques to study, design, and optimize nanomaterials. The content is organized to guide you from foundational concepts to advanced applications and algorithm implementations through example codes.

The book begins with an overview of the principles of molecular simulation and the basics of machine learning, ensuring a common understanding regardless of your background. The subsequent chapters delve into specific Machine Learning (ML) techniques—such as ensemble methods and kernel methods—and how they are used to enhance the accuracy, efficiency, and scalability of simulations at the nanoscale. You will also discover practical strategies for data generation, feature engineering, model training, and validation, with a strong emphasis on physical interpretability and domain relevance. The practical implementation of cutting-edge research is included throughout to illustrate the design of nanomaterials for multi-faceted applications, covering low- and high-dimensional materials. Whether you are a researcher, student, or practitioner, this book equips you with the theoretical insight and practical tools needed to drive innovation at the intersection of nanotechnology, computational science, and artificial intelligence.

More specifically, Chap. 1 provides a general introduction to highlight the motivation for considering nanoengineered materials in a range of engineering applications. Chapter 2 offers an overview of molecular dynamics simulations, including various interatomic potentials and the modelling of nanoscale materials. Chapter 3 introduces machine learning, discussing both parametric and non-parametric models, along with a general pipeline. Chapter 4 reviews the prospects of machine learning-driven atomistic simulations, focusing on challenges and opportunities in high-throughput simulations and nanoscale design. Chapter 5 explores the fracture response of graphene using a data-driven characterization approach, involving machine learning pipelines. Chapter 6 addresses the nanoscale ballistic response of bi-layer graphene with a machine learning-driven approach. Chapter 7 presents interatomic potential parametrization for graphene, emphasizing the role of machine learning in feature importance and parameter modifications. Chapter 8 discusses high-entropy alloys (HEAs) with a data-driven quasi-static characterization and machine learning deployment for functional design. Chapter 9 focuses on the optimal nanoscale ballistic performance of HEAs, utilizing machine learning for Bayesian optimization. Finally, Chap. 10 delves into materials informatics, covering dataset generation, scalability of machine learning models, and domain-specific awareness in model development. The scope of this book includes the aspect of molecular dynamics modelling as well as a critical evaluation of the efficient machine learning-based approaches for nanomaterials.

Acknowledgements: The authors are deeply indebted to numerous colleagues, students, collaborators, and mentors for stimulating scientific discussions, exchanges of ideas, and on many occasions direct contributions towards the intellectual content of the book. The authors would like to gratefully acknowledge the contribution of Mr. Subrata Barman (NIT Silchar), Dr. Gautam (Dayanand Sagar University, India), Dr. Subhankar Saha (Brainware University), Dr. Aditya Roy (NIT Silchar), Dr. Lintu Roy (NIT Silchar), Prof. Susmita Naskar (University of Southampton, UK), and Prof. Sondipon Adhikari (University of Glasgow, UK). Besides the names taken here, we are thankful to many colleagues, fellow researchers, and students working in this field of research around the world, whose names cannot be listed here due to page limitations. The lack of explicit mentions by no means implies that their contributions are any less. The opinions presented in the book are entirely ours, and none of our colleagues, students, collaborators, and mentors has any responsibility for any shortcomings. We have been fortunate to receive grants from agencies like DRDO, ISRO, SERB, and SPARC, Government of India. Without such funding, it would have been impossible to conduct the research leading to this book.

We welcome suggestions and feedback from our readers.

Coimbatore, India — Kritesh Kumar Gupta
g_kriteshkumar@cb.amrita.edu

Silchar, India — Sudip Dey
sudip@mech.nits.ac.in

Southampton, UK — Tanmoy Mukhopadhyay
t.mukhopadhyay@soton.ac.uk

Contents

About the Authors

Dr. Kritesh Kumar Gupta is an Assistant Professor at the School of Artificial Intelligence, Amrita Vishwa Vidyapeetham, Coimbatore, India. He holds a Ph.D. (specialising in Machine Learning Driven Nanoscale Design of Materials) in Engineering from the National Institute of Technology Silchar, India. Dr. Gupta specializes in developing cutting-edge computational frameworks that integrate atomistic simulations with artificial intelligence frameworks to accelerate materials design. His expertise in advanced machine learning and deep learning techniques is pivotal in advancing materials design and discovery. He has contributed significantly to the field through high-impact publications and presentations on national and international platforms. He has authored/coauthored 25+ journal papers, and 10+ conference papers. He is an associate member of professional bodies such as the Institution of Engineers, India (IEI), and the Institute of Physics (IOP). His current research focuses on deploying the pioneering concepts in AI-driven materials science, particularly in designing nano-engineered function-specific material systems.

Dr. Sudip Dey is an Associate Professor at the Mechanical Engineering Department of the National Institute of Technology Silchar, India. He holds a Bachelor's degree in Mechanical Engineering from Jadavpur University, India, and a Ph.D. in Engineering from the same institution. Dr. Dey's expertise in Mechanics, Design, and Materials has been shaped by his international experiences, including post-doctoral research at Leibniz-Institut für Polymerforschung Dresden e. V., Germany, and Swansea University, United Kingdom. With over 20 years of experience in research, teaching, and industrial endeavours, Dr. Dey is a leading figure in Uncertainty Quantification (UQ) and has authored the influential book *Uncertainty Quantification in Laminated Composites: A Meta-model Based Approach*. His research interests span Molecular Dynamics, Tribology, Metamaterials, and Digital Twin, contributing significantly to the field through publications in prestigious international journals and conferences.

Dr. Tanmoy Mukhopadhyay is appointed as a Senior Lecturer (LB) and leads the Programmable Matter Lab at the University of Southampton, UK. His academic journey is marked by notable positions at esteemed academic institutions worldwide.

Before joining the University of Southampton, Dr. Mukhopadhyay held a role as an Assistant Professor in the Aerospace Engineering Department of IIT Kanpur. Prior to that, he made significant contributions to the field of metamaterial intelligence as a postdoctoral research fellow at the University of Oxford, UK. Dr. Mukhopadhyay completed his Ph.D. at Swansea University, UK, where his exceptional research was recognised through the award of Best Ph.D. in Engineering (Gold Medal). His research interests and expertise broadly lie in the field of mechanics and multi-physics analysis, focusing on mechanical metamaterials and advanced functional composites at multiple length scales that involve cutting-edge developments at the intersection of artificial intelligence, advanced physics-informed simulation and materials science. He is actively involved in shaping the research directions in Aerospace and Mechanical Engineering through his contributions as an editorial board member of multiple prestigious journals like *Artificial Intelligence Review*, *Space: Science and Technology*, *Scientific Reports*, *Discover Applied Sciences*, *Reviews on Advanced Materials Science*, *Discover Materials*, *Journal of Composites Science*, *Frontiers in Nanotechnology* etc. along with multiple international conferences.

Chapter 1
Introduction to Nanoengineered Materials

Atoms are very special: they like certain particular partners, certain particular directions, and so on. It is the job of physics to analyze why each one wants what it wants.
—Richard P. Feynman

1.1 Overview

This quote by Richard Feynman, along with his seminal 1959 talk 'There's Plenty of Room at the Bottom', delivered at the American Physical Society at Caltech, introduced the groundbreaking concept of manipulating matter at the atomic scale. The quote highlights the role of physics-based characterization in understanding atomically engineered materials. However, in today's era of rapid technological advancement, data-driven approaches are increasingly replacing and augmenting traditional physics-based methods. Motivated by this shift, the following chapters in this book explore machine learning-driven nanoscale design for two distinct classes of materials: low-dimensional materials such as graphene, and high-dimensional materials like high-entropy alloys. The operational performance of material systems at the macro scale is the culmination of its structural attributes at the atomic-scale structural changes. This is certainly relevant for heterogeneously engineered material systems like composite materials, metal alloys, and so on. As a result, investigating material behaviour at the atomic scale aids in revealing the complex physical mechanism underlying a specific phenomenon occurring at the macro scale. In this context, this chapter lays the foundation for the entire book by discussing the several challenges associated with atomistic simulations, and, finally, highlighting the rationale for integrating machine learning (ML) with atomistic simulations.

When it comes to the analysis of nanoscale material, nanomechanical modelling is proven to be a more suitable procedure rather than nanoscale experimentation [1]. Setting up the nanoscale experimentation has its limitations in terms of length and time scale (refer to Fig. 1.1); apart from that, it is a tedious and expensive task to perform. There are mainly two ways in which nanomechanical modelling could

K. K. Gupta et al., *Machine Learning in Nanoscale Materials Design*, Materials Horizons: From Nature to Nanomaterials, https://doi.org/10.1007/978-981-95-2660-4_1

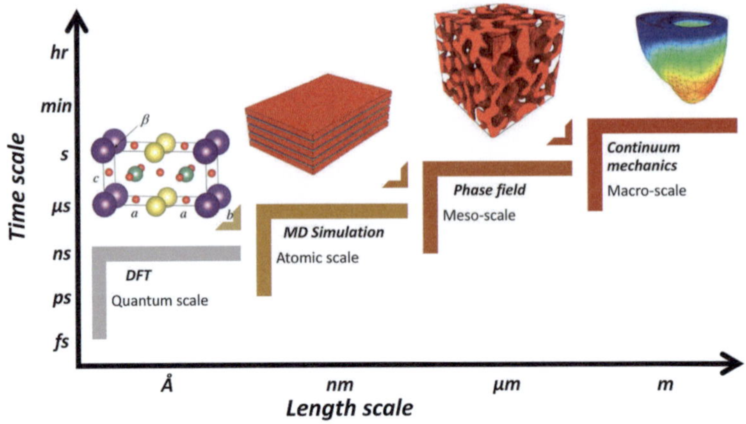

Fig. 1.1 Time and length scale variations in computational exercises

be carried out: one is the *first principles* approach, and another is a semi-empirical approach. The *first principles* approach or high fidelity ab initio *methods* such as Density Functional Theory (DFT) and Hartree–Fock method are more accurate than the semi-empirical method such as molecular dynamics (MD) simulation, as these methods solve the fundamental equation (*i.e.* Schrodinger equation) for the whole system with minimal approximations; however, the first principles approach has the limitation over the size of the system and the computational expense [2]. On the other hand, the semi-empirical method such as molecular dynamics (MD) simulation is more computationally affordable and can deal with larger (high number of atoms) systems than the *first principles* approaches.

Recent advancements in computational capabilities, combined with the time and length scale constraints in experimentally exploring materials at smaller dimensions, highlight the pressing need to computationally analyse material behaviour [3]. While computational methods like MD simulations provide an effective platform for nanoengineering material systems, there are a few challenges associated with these approaches, as follows.

1.1.1 Computationally Intensive and Expensive Nature

Experimental analysis of nanoscale materials is limited by factors such as precision, cost, and time [4]. With advancements in computational capabilities, researchers are increasingly turning to simulations as a partial substitute for experimental methods, particularly in nanoscale analysis. This shift has led to a significant rise in the use of molecular dynamics (MD) simulations by researchers over the past few decades [5]. One of the major limitations of these simulations is that they remain computationally expensive and time-consuming, making it challenging to thoroughly analyse a nanoscale system. High-throughput simulations require supercomputing resources and still take considerable time to process larger systems (> 10,000 atoms) [6]. This slows down the efficient discovery of materials and the ability to gain comprehensive computational insights into their mechanical and other multi-physical behaviours.

The recent advances in machine learning have paved new paths in nanoscale materials research, enabling the efficient exploration of the limits of various multi-physical properties with the help of artificial intelligence [7, 8]. The collaboration between machine learning and computational materials science has elevated the nanoscale characterization of materials to a whole new level. The generalization and prediction capabilities provided by advanced machine learning frameworks act as a surrogate, enabling large-scale predictions for virtually generated samples [9–14].

1.1.2 Interatomic Potentials (IPs)

MD simulation is considered a computational experimentation procedure that is used to understand the behaviour of nanomaterials in the desired physical conditions. The observations obtained from MD simulations greatly depend on the interatomic potential (IP) used for the specific analysis [15]. These IPs are calibrated with the experimental data or the data received from DFT simulations to mimic the desired properties of the system under consideration. When the simulations are conducted for the observations beyond the calibrated envelope of these IPs, the validity of predicted results is questionable. The lack of suitable interatomic potentials (IPs) for custom-designed nanomaterials presents a challenge in performing MD simulations. In this regard, machine learning methods can be combined with DFT calculations to create novel IPs for the targeted material system. The machine learning-based interatomic potential (MLIP) is trained to establish the relationship between interatomic distance and the potential energy of the material system [16]. This is another avenue in atomistic simulations where machine learning has made significant and successful contributions.

1.2 Atomistic Simulations in the Era of Materials Informatics

The introduction of new materials propels civilizations forward, to the point where various periods of human history are frequently defined by the materials used in technology, such as the Stone, Bronze, and Iron Ages, and, more recently, the Silicon or Polymer Age. The design and discovery of novel classes of materials are crucial to technological growth because they make practical processes more efficient, dependable, and sustainable, allowing for breakthroughs that can handle difficult challenges across multiple industries [17]. For instance, the development of biocompatible polymers and advanced ceramics has enabled advancements in targeted drug delivery systems and orthopaedic implants, resulting in substantial therapeutic breakthroughs

[18–21]. Likewise, the discovery of lithium-based compounds [22] and perovskite materials [23] has facilitated the development of high-performance batteries and solar cells. Throughout history, military demands have frequently spurred material innovation, as seen in the development of advanced materials for weapons and protective equipment. The progress of composite materials and advancements in metallurgy can be linked to the need for enhanced performance in military applications [24]. This highlights the multi-faceted nature of material development, where the pursuit of innovations in health, sustainability, security, and efficiency fuels the search for new materials. The current quest for room-temperature superconductors remains a prominent area of focus, as these materials promise to eliminate the energy losses associated with electrical resistance. The successful development of room-temperature superconductors could substantially revolutionize another ongoing quest for quantum computing by eliminating the need for expensive and complex cooling systems, qubit stability, faster quantum operations, etc. [25]. With the recent technological advancements in computational capabilities, materials design and discovery have taken a route of artificial intelligence (AI)-driven materials discovery. The rapid growth of machine learning (ML) and deep learning (DL) frameworks for materials characterization has transformed how materials are designed, discovered, and studied.

Traditionally, materials discovery relied heavily on trial-and-error experiments, which were not only time-consuming but also resource-intensive. Moreover, these methods often resulted in an Edisonian approach, where discoveries were made only after numerous failed attempts. This is due to the intricate complexity of the process of materials design and a vast number of potential combinations arising from the inherent relationships between processing, structure, properties, and performance [26]. This established a strong rationale for devising a robust mechanism for materials design by harnessing the computationally efficient frameworks under the purview of the materials genome initiative (MGI) [27]. The schematic of the materials design pipeline for the traditional approach versus materials informatics is illustrated in Fig. 1.2. The concept of materials informatics facilitates the effective combination of computational atomistic simulations (requiring a minimal number of samples) with AI-driven frameworks to strategically explore the material space for the optimal solution, followed by prototype synthesis and characterization. This eliminates the synthesis, fabrication, and characterization costs typically associated with the traditional materials screening process.

1.3 Nanomaterials Explored in the Book

This book aims to provide an applied understanding of machine learning-driven molecular dynamics simulations for two distinct classes of materials: low-dimensional and high-dimensional. The machine learning-based characterization of a representative low-dimensional material, such as graphene, is explored under both

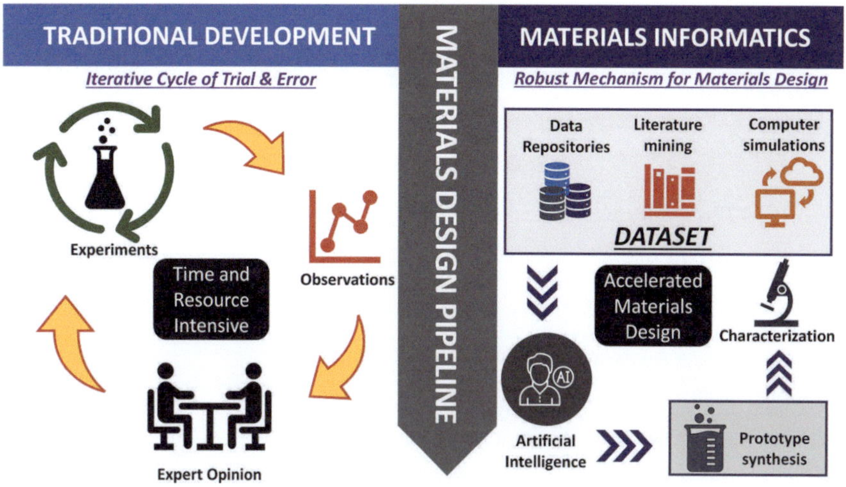

Fig. 1.2 The common materials design pipeline utilized in the traditional approach and materials informatics

quasi-static and dynamic loading conditions. Similarly, for high-dimensional materials, the book presents the machine learning-driven design of high-entropy alloys (HEAs) as a representative case study.

1.3.1 Graphene

The nanoscale allotrope of carbon came into existence with the discovery of fullerene in 1985 by Kroto et al. [28], which further led to the discovery of carbon nanotubes (CNTs) by Iijima [29] in 1991. By this time only 3D (graphite and diamond), 1D (CNTs), and 0D (fullerenes) allotropes were known to the science community. It was realized in 1991 that CNTs were formed by rolling carbon 2D material, following which in 2004 first time graphene was extracted from the graphite using the scotch tape method by Andre Geim [30]. Since the invention of graphene, a series of 2D materials and their variants have been extensively explored by the materials research community [31–34]. Graphene is building a block of some of the carbon allotropes such as graphite, fullerenes, and CNTs. Graphene is a monolayer of carbon atoms arranged in a hexagonal lattice (refer to Fig. 1.3).

Graphene has the sp^2 bonded carbon atoms with the densely packed honeycomb crystal lattice. Its extraordinary properties such as low density, high stiffness, high fracture strength, and excellent electrical and thermal conductivity make it a suitable candidate for a wide range of applications. So far, a vast amount of research has been carried out in the past to find the exact mechanical properties of graphene. The exceptional properties of graphene have been utilized to develop a

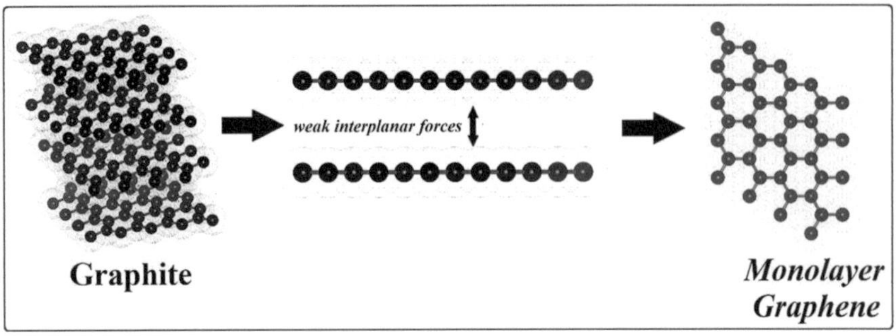

Fig. 1.3 Monolayer graphene: a building block of graphite [31]

novel class of nanocomposites, nanoactuators, nano-resonators, biosensors, biomechanical devices, and nanoelectromechanical system (NEMS) applications [35–39]. For several years, the multi-physical properties of graphene have piqued the interest of the research community. For instance, Datta et al. [40] proposed the utility of defective graphene as the anode material for sodium and calcium ion batteries. Chowdhury et al. [41] reported the variation in the transmission of graphene nanoribbons when exposed to the bio-molecules; therefore, it has a promising application in bio-sensing. Kommu and Singh [42] demonstrated the capability of separating pollutants from water by using graphene and hBN-based heterostructures; similarly in another study by Kommu et al. [43], graphene is used to remove the heavy metal ions from the water by utilizing functionalized graphene membranes. Such studies carried out in the past highlight the multi-physical capabilities of graphene and therefore its multi-faceted applications. Figure 1.4 depicts a few domain-specific applications of graphene, along with the physical properties of graphene that are utilized in the application. It is anticipated that extreme mechanical strength and stiffness can be harnessed to produce hybrid lightweight material systems for various structural applications.

1.3.2 High-Entropy Alloys (HEAs)

High-entropy alloys (HEAs) are a class of multi-principal element alloys comprised of five or more principal alloying elements, each typically present in atomic percentages ranging from 5 to 35% [44, 45]. This unique feature of HEAs in terms of wide compositional space (multiple principal elements) promotes mutual solubility and enhanced configurational entropy. The mutual solubility and high configurational entropy further ensure the formation of solid solutions with the presence of uniform phases. In contrast to the expectations of complex phase formation in multi-principal element alloys, high-entropy alloys (HEAs) typically exhibit relatively simple phases, primarily consisting of face-centred cubic (FCC), body-centred cubic (BCC), or a combination of both. This structural simplicity is responsible for the exceptional properties of HEAs [46, 47], including robust cryogenic toughness, high strength,

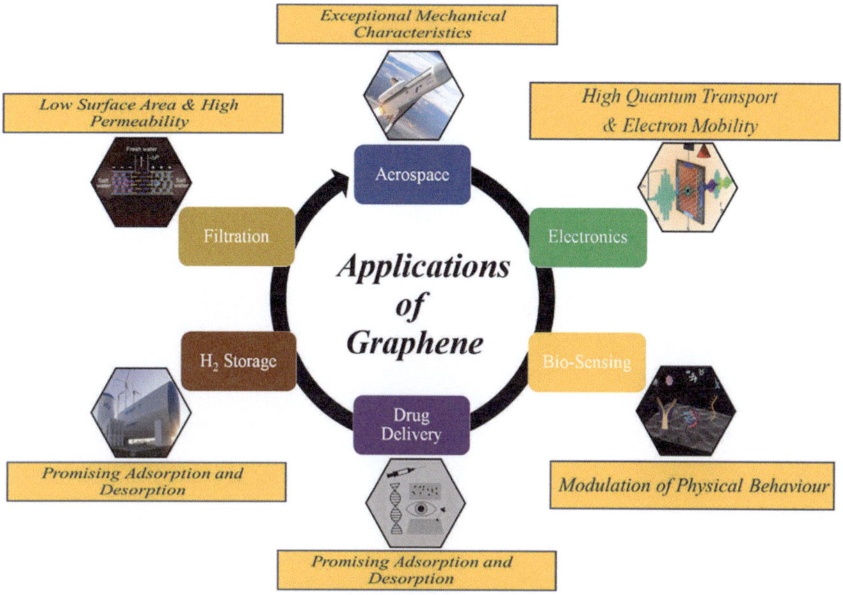

Fig. 1.4 Domain-specific applications of graphene

thermal stability at elevated temperatures, and excellent resistance to corrosion and wear. Such remarkable mechanical characteristics of HEAs have piqued the interest of the materials science community in harnessing their exceptional performance in multi-faceted applications (refer to Fig. 1.5).

1.4 Organization of the Book

Hereafter, the book is organized as follows:

Chapter 2 introduces the fundamental aspects of molecular dynamics (MD) simulations, focusing on the LAMMPS environment. It clearly outlines the essential components required to set up and perform simulations and provides a step-by-step guide for executing these simulations on local machines. Additionally, it presents the input scripts used in the case studies discussed in Chaps. 5 through 9.

Chapter 3 offers a comprehensive overview of various classes of machine learning (ML) models that can be integrated with atomistic simulations. It outlines the operational principles of the machine learning (ML) algorithms utilized in the subsequent chapters of this book. A detailed conceptual overview is provided for some of the more technologically advanced ML models, including ensemble learning techniques (such as bagging and boosting) and Gaussian Process Regression (GPR). The chapter also introduces a standardized ML pipeline tailored for ML-driven atomistic design of materials.

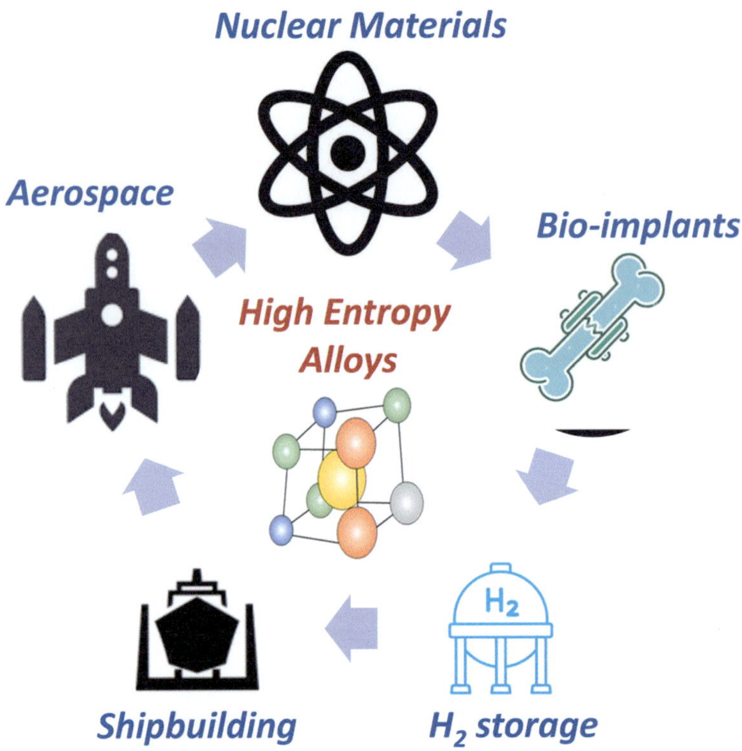

Fig. 1.5 Domain-specific applications of high-entropy alloys (HEAs)

Chapter 4 provides a comprehensive review of machine learning-integrated molecular dynamics (MD) simulations, emphasizing the potential of ML to address fundamental challenges in atomistic simulations.

Chapters 5 focuses on the applied aspects of materials informatics (ML-driven MD simulations), showcasing the implementation of ML-based predictive pipelines for materials design and optimization. This chapter develops an ML-driven framework to characterize the fracture behaviour of different graphene configurations.

Chapter 6 deploys the ML regression and classification framework for characterizing the nanoscale ballistic performance of the bilayer graphene.

Chapter 7 builds upon Chap. 5 by leveraging the generalization capabilities of ML models to parameterize the Tersoff interatomic potential, ensuring that MD simulation outcomes closely align with experimental observations and thus yield high-fidelity results.

Chapter 8 presents the integration of an ML-based predictive framework with a multi-objective optimization strategy to design AlCoCrFeNi high-entropy alloys (HEAs) for a high stiffness-to-weight ratio. This approach enables the exploration of Pareto-optimal compositions that simultaneously enhance stiffness and reduce material density.

Chapter 9 extends this work by incorporating the predictive pipeline into a Bayesian optimization framework, aimed at optimizing the composition of AlCoCr-FeNi HEAs for improved kinetic energy dissipation under high-velocity impact scenarios.

Chapter 10 discusses current trends in complementing atomistic simulations with materials informatics strategies to reduce the cost and resource intensiveness of the initial materials screening process.

References

1. Barman S, Kumar Gupta K, Dey S (2024) Exploring nano-scale scratching induced tribological behavior of graphene engineered AlCoCrFeNi high-entropy alloy. J Appl Mech 91(10)
2. Gupta KK, Mukhopadhyay T, Roy A, Dey S (2020) Probing the compound effect of spatially varying intrinsic defects and doping on mechanical properties of hybrid graphene monolayers. J Mater Sci Technol 50:44–58
3. Pilania G, Goldsmith BR, Yoon M, Dongare AM (2022) Recent advances in computational materials design: methods, applications, algorithms, and informatics. J Mater Sci 57(23):10471–10474
4. Parak WJ, Nel AE, Weiss PS (2015) Grand challenges for nanoscience and nanotechnology. ACS Nano 9(7):6637–6640
5. Taha TB, Barzinjy AA, Hussain FHS, Nurtayeva T (2022) Nanotechnology and computer science: trends and advances. Memories-Mater Dev Circ Syst 2:100011
6. Afzal MAF, Browning AR, Goldberg A, Halls MD, Gavartin JL, Morisato T, Hughes TF, Giesen DJ, Goose JE (2020) High-throughput molecular dynamics simulations and validation of thermophysical properties of polymers for various applications. ACS Appl Polym Mater 3(2):620–630
7. Venkatraman A, Wilson MA, Montes de Oca Zapiain D (2025) Accelerating charge estimation in molecular dynamics simulations using physics-informed neural networks: corrosion applications. NPJ Comput Mater 11(1):24
8. Gupta KK, Mukhopadhyay T, Roy A, Roy L, Dey S (2021) Sparse machine learning assisted deep computational insights on the mechanical properties of graphene with intrinsic defects and doping. J Phys Chem Solids 155:110111
9. Gupta KK, Roy L, Dey S (2022) Machine learning-based molecular dynamics simulations of monolayered graphene. In: Recent advances in computational and experimental mechanics, Vol II: Select proceedings of ICRACEM 2020. Springer Nature Singapore, Singapore, pp 251–263
10. Gupta KK, Mukhopadhyay T, Roy L, Dey S (2022) High-velocity ballistics of twisted bilayer graphene under stochastic disorder. Adv Nano Res 12(5):529–547
11. Roy A, Gupta KK, Dey S (2022) Probabilistic investigation of temperature-dependent vibrational behavior of hetero-nanotubes. Appl Nanosci 12(7):2077–2089
12. Gupta KK, Roy L, Dey S (2022) Hybrid machine-learning-assisted stochastic nano-indentation behaviour of twisted bilayer graphene. J Phys Chem Solids 167:110711
13. Gupta KK, Roy A, Mukhopadhyay T, Roy L, Dey S (2022) Probing the stochastic fracture behavior of twisted bilayer graphene: efficient ANN based molecular dynamics simulations for complete probabilistic characterization. Mater Today Commun 32:103932
14. Gupta KK, Roy L, Dey S (2022) Ballistic performance of bi-layer graphene: artificial neural network based molecular dynamics simulations. In: Machine learning applied to composite materials. Springer Nature Singapore, Singapore, pp 135–148
15. Gupta KK, Mukhopadhyay T, Roy L, Dey S (2022) Hybrid machine-learning-assisted quantification of the compound internal and external uncertainties of graphene: towards inclusive analysis and design. Mater Adv 3(2):1160–1181

16. Wang G, Wang C, Zhang X, Li Z, Zhou J, Sun Z (2024) Machine learning interatomic potential: bridge the gap between small-scale models and realistic device-scale simulations. Iscience
17. Rogers JA, DeSimone JM (2016) Novel materials. Proc Natl Acad Sci 113(42):11667–11669
18. Hakim LK, Yazdanian M, Alam M, Abbasi K, Tebyaniyan H, Tahmasebi E, Khayatan D, Seifalian A, Ranjbar R, Yazdanian A (2021) Biocompatible and biomaterials application in drug delivery system in oral cavity. Evidence-Based Complemen Alt Med 2021(1):9011226
19. Zou Y, Huang B, Cao L, Deng Y, Su J (2021) Tailored mesoporous inorganic biomaterials: assembly, functionalization, and drug delivery engineering. Adv Mater 33(2):2005215
20. Yadav D, Garg RK, Ahlawat A, Chhabra D (2020) 3D printable biomaterials for orthopedic implants: solution for sustainable and circular economy. Resour Policy 68:101767
21. Filip N, Radu I, Veliceasa B, Filip C, Pertea M, Clim A, Pinzariu AC, Drochioi IC, Hilitanu RL, Serban IL (2022) Biomaterials in orthopedic devices: current issues and future perspectives. Coatings 12(10):1544
22. Wietelmann U, Klett J (2018) 200 years of lithium and 100 years of organolithium chemistry. Z Anorg Allg Chem 644(4):194–204
23. Katz EA (2020) Perovskite: name puzzle and German-Russian odyssey of discovery. Helv Chim Acta 103(6):e2000061
24. Wellmann PJ (2021) The search for new materials and the role of novel processing routes. Discov Mater 1(1):14
25. Bravyi S, Dial O, Gambetta JM, Gil D, Nazario Z (2022) The future of quantum computing with superconducting qubits. J Appl Phys 132(16)
26. Choudhary K, DeCost B, Chen C, Jain A, Tavazza F, Cohn R, Park CW, Choudhary A, Agrawal A, Billinge SJ, Holm E (2022) Recent advances and applications of deep learning methods in materials science. NPJ Comput Mater 8(1):59
27. de Pablo JJ, Jackson NE, Webb MA, Chen LQ, Moore JE, Morgan D, Jacobs R, Pollock T, Schlom DG, Toberer ES, Analytis J (2019) New frontiers for the materials genome initiative. NPJ Comput Mater 5(1):41
28. Kroto HW, Heath JR, O'Brien SC, Curl RF, Smalley RE (1985) C60: Buckminsterfullerene. Nature 318(6042):162–163
29. Iijima S (1991) Helical microtubules of graphitic carbon. Nature 354(6348):56–58
30. Geim AK, Novoselov KS (2010) The rise of graphene. In: Nanoscience and technology: a collection of reviews from nature journals, pp 11–19
31. Saumya K, Naskar S, Mukhopadhyay T (2023) 'Magic' of twisted multi-layered graphene and 2D nano-heterostructures. Nano Futures 7(3):032005
32. Mahata A, Mukhopadhyay T (2018) Probing the chirality-dependent elastic properties and crack propagation behavior of single and bilayer stanene. Phys Chem Chem Phys 20(35):22768–22782
33. Mukhopadhyay T, Mahata A, Adhikari S, Zaeem MA (2017) Effective mechanical properties of multilayer nano-heterostructures. Sci Rep 7(1):15818
34. Mukhopadhyay T, Mahata A, Adhikari S, Zaeem MA (2017) Effective elastic properties of two dimensional multiplanar hexagonal nanostructures. 2D Mater 4(2):025006
35. Mukhopadhyay T, Mahata A, Adhikari S, Zaeem MA (2018) Probing the shear modulus of two-dimensional multiplanar nanostructures and heterostructures. Nanoscale 10(11):5280–5294
36. Roy A, Gupta KK, Dey S, Mukhopadhyay T (2024) Random topological defects in double-walled carbon nanotubes: on characterization and programmable defect-engineering of spatio-mechanical properties. Adv Nano Res 16(1):91–109
37. Muthukrishnaraj A, Kalaivani SS, Suresh R, Manikandan A, Arularasu MV, Tamilnayagam V (2025) Diagnostic applications of magnetic nanohybrids. In: Functionalized magnetic nanohybrids. Elsevier, pp 169–200
38. Gupta KK, Dey S (2019) Effect of temperature on the fracture strength of perfect and defective monolayered graphene. In: Advances in computational methods in manufacturing: select papers from ICCMM 2019. Springer Singapore, pp 793–804
39. Selvam SK, Kumar SRA, Balasubramanian N, Kumar PS, Muthukrishnaraj A, Kalaivani SS, Vinayagam R, Azarudeen RS, Ahamed MAR, Harikrishnan A, Josephine GS (2024) Recent

advancements in the environmental significance of the graphitic carbon nitride-based materials. Desalin Water Treat:100385

40. Datta D, Li J, Shenoy VB (2014) Defective graphene as a high-capacity anode material for Na-and Ca-ion batteries. ACS Appl Mater Interfaces 6(3):1788–1795
41. Chowdhury R, Adhikari S, Rees P, Wilks SP, Scarpa F (2011) Graphene-based biosensor using transport properties. Phys Rev B 83(4):045401
42. Kommu A, Singh JK (2017) Separation of ethanol and water using graphene and hexagonal boron nitride slit pores: a molecular dynamics study. J Phys Chem C 121(14):7867–7880
43. Kommu A, Namsani S, Singh JK (2016) Removal of heavy metal ions using functionalized graphene membranes: a molecular dynamics study. RSC Adv 6(68):63190–63199
44. Barman S, Gupta KK, Dey S (2024) Probing atomistic deformation behavior of graphene-coated Al0. 3CoCrFeNi high-entropy alloy under nanoindentation. J Micromech Mol Phys 9(01):1–14
45. Barman S, Gupta KK, Dey S (2024) Enhancing mechanical performance of Al0. 3CoCrFeNi HEA films through graphene coating: insights from nanoindentation and dislocation mechanism analysis. Modell Simul Mater Sci Eng 32(3):035012
46. Gupta KK, Barman S, Dey S, Naskar S, Mukhopadhyay T (2024) On exploiting nonparametric kernel-based probabilistic machine learning over the large compositional space of high entropy alloys for optimal nanoscale ballistics. Sci Rep 14(1):16795
47. Gupta KK, Barman S, Dey S, Mukhopadhyay T (2024) Explainable machine learning assisted molecular-level insights for enhanced specific stiffness exploiting the large compositional space of AlCoCrFeNi high entropy alloys. Mach Learn: Sci Technol 5(2):025082

Chapter 2
Molecular Dynamics Simulation: An Overview

2.1 Overview

Molecular dynamics (MD) is a computational technique used to analyse the physical behaviour of material systems at the atomic scale. In an MD approach, the material system is considered as a group of rigid bodies (atoms), wherein the dynamic evolution of atoms is evaluated by numerically solving Newton's equations of motion. This method is commonly used in biophysics, chemistry, and materials science to understand the fundamental causes of a system's behaviour at the macro-molecular state. In this approach, semi-empirical models are used to simulate the interatomic interactions between constituent atoms. These semi-empirical models define the interatomic forces, which are then used to calculate the potential energy of the material system under consideration. The design constraints in the MD approach are enforced by utilizing the statistical ensembles, such as NPT, NVT, and NVE. The MD method utilizes a sequential approach to perform nanoscale investigation, wherein at first the initial state of the system is defined in terms of positional coordinates of the constituent atoms, then the interatomic interactions are defined by utilizing the semi-empirical potential functions, and, at last, the trajectory of the atoms is predicted by integrating Newton's equations of motion. This chapter focuses on the key practical aspects of performing molecular dynamics (MD) simulations of graphene and high-entropy alloys. These two classes of materials are emphasized because they are investigated in detail in the subsequent chapters.

The molecular dynamics (MD) simulation is considered one of the best alternatives for simulating nanoscale experiments which are difficult to perform in a laboratory setting. The MD considers the materials as the collection of individual atoms. The mutual interaction among the constituent atoms is enforced by empirically formulated potential function, which leads to the determination of interatomic forces. The constituent atoms further follow Newton's law of motion, and the trajectory of the atoms is captured with the help of numerical integration. In simple terms, MD is an integration of motion with respect to time to determine the growth of the

K. K. Gupta et al., *Machine Learning in Nanoscale Materials Design*, Materials Horizons: From Nature to Nanomaterials, https://doi.org/10.1007/978-981-95-2660-4_2

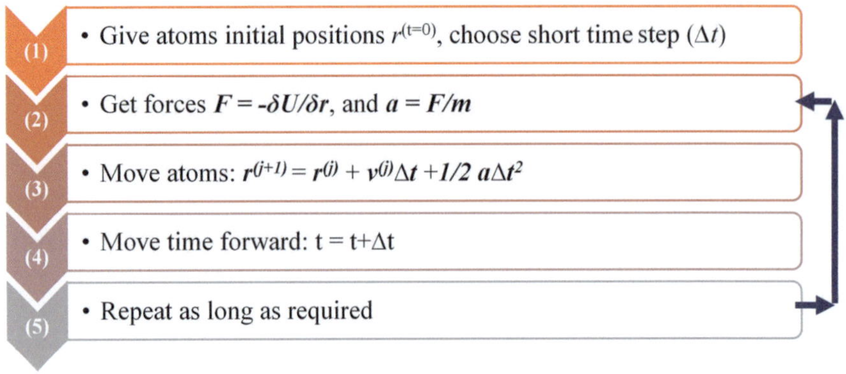

Fig. 2.1 Steps involved in MD simulations

system with time. It is a common practice to initiate MD simulation by performing the energy minimization of the modelled material system.

The steps incurred in the energy minimization are as follows (refer to Fig. 2.1) [1]:

- With the provided initial position and velocities of the constituent atoms and the potential field used, the interatomic forces on each atom are determined.
- With the obtained information the system of atoms is progressed towards the lower energy states through a small-time interval, which leads to new positions and velocities of the atoms.
- The obtained new data of positions and velocities are considered as the new input and the same process is repeated over thousands of small time steps until the equilibrium is established and the properties of the system do not change with time.

The desired simulations are then performed on the obtained structure (in a minimum potential energy state) by enforcing the necessary boundary conditions. The ingenuity of observations drawn from the MD simulations greatly depends on the selection of appropriate interatomic potential. When an individual atom of the material system reaches the vicinity of other atoms, the equilibrium condition between attraction and repulsion comes into play, and it is determined by interatomic potential. Atoms further follow Newton's law of motion and tend to obtain the minimum potential states. The summation of all the forces acting on an atom, F, is defined as

$$F = ma = m\frac{\mathrm{d}v}{\mathrm{d}t} = m\frac{\mathrm{d}^2 r}{\mathrm{d}t^2} \tag{2.1}$$

where m denotes the atomic mass, a denotes the acceleration, v stands for the velocity of an atom, and r denotes the position of an individual atom. With total energy E to be constant ($\mathrm{d}E/\mathrm{d}t = 0$), F is directly proportional to the negative gradient of potential with respect to the position of constituent atoms:

$$F = -\Delta U \rightarrow F = -\Delta U(r_1, r_2, \ldots, r_n) \tag{2.2}$$

where U refers to the potential energy and r_1, r_2, ..., r_n denote the position of constituent atoms. This clarifies that if the potential energy of the system is known with respect to the interatomic distance, then the interatomic forces may be determined and thus the above equation can be solved for the time evolution of the system; and once the interatomic forces are known, the acceleration of the atoms may be determined using Newton's second law of motion. Further, the system is allowed to progress with the obtained acceleration for a certain time period known as time step (generally in the order of femtoseconds, 10^{-15} s). At the end of each time step, the new position and velocity of the atoms are obtained using the velocity Verlet algorithm. The velocity Verlet algorithm may be defined as

$$v\left(t_o + \frac{\Delta t}{2}\right) = v(t_o) + a(t_o)\frac{\Delta t}{2} \tag{2.3}$$

$$r(t_o + \Delta t) = r(t_o) + v\left(t_o + \frac{\Delta t}{2}\right)\Delta t \tag{2.4}$$

$$v(t_o + \Delta t) = v\left(t_o + \frac{\Delta t}{2}\right) + a(t_o)\frac{\Delta t}{2} \tag{2.5}$$

where t_o is the initial time and Δt is the time step.

The generation of interatomic potential for a specific system involves fitting certain functions with the parameters of experimental data or data received through *first-principles* methods. The data used for fitting mostly is inclusive of vibration spectrum, thermal expansion coefficient, elastic modulus, equilibrium lattice parameter, vacancy formation energies, cohesive energy, etc. This must be kept in mind that these interatomic potentials are system-specific [1]. The following subsections provide a detailed discussion of key practical aspects involved in performing molecular dynamics (MD) simulations. These include the mathematical foundations of commonly used interatomic potentials—such as Tersoff and AIREBO for nanoscale characterization of graphene, and EAM/Alloy for high-entropy alloys—as well as methods for controlling temperature and pressure within the MD environment. Additionally, the role and implementation of various statistical ensembles are explored.

2.1.1 Tersoff Interatomic Potential

The fracture behaviour of the graphene structures examined in Chaps. 5 and 7 is simulated using the Tersoff interatomic potential. Similarly, the silicon projectile discussed in Chap. 6 is also modelled with the Tersoff potential [2–4]. The Tersoff potential is a three-body model that captures the interactions between atoms, with the energy E between atoms defined as

$$E = \frac{1}{2}\sum_{i}\sum_{j \neq i} V_{ij} \tag{2.6}$$

where

$$V_{ij} = f_C\left(r_{ij}\right)\left[f_R(r_{ij}) + b_{ij} f_A\left(r_{ij}\right)\right] \tag{2.7}$$

and V_{ij} represents the potential energy of the system, while f_R and f_A denote the repulsive and attractive pair potentials, respectively, and f_C is the cut-off function. This formula accounts for the potential energy from all neighbouring atoms (j and k) of atom i within the specified cut-off radius. The separation between adjacent atoms i and j is denoted by r_{ij}. The parameter b_{ij} is an empirical bond-order coefficient, with f_R and f_A described as

$$f_R = Ae^{-\lambda_1} \tag{2.8}$$

and

$$f_A = Be^{-\lambda_2} \tag{2.9}$$

where other parameters used for two-body interactions are A, B, λ_1, and λ_2 [5].

2.1.2 AIREBO-Morse Interatomic Potential

The interactions between carbon atoms within both layers of bilayer graphene (BLG), as discussed in Chap. 6, are simulated using the AIREBO-Morse force field [6]. This force field is a modified version of the standard AIREBO model, in which the traditional Lennard–Jones potential is replaced by a Morse potential. This adjustment allows for a more accurate representation of bond deformation and breakage under high-pressure conditions [7, 8]. The AIREBO-Morse potential is mathematically expressed as

$$E_{\text{AIREBO - M}} = \frac{1}{2}\sum_{j}\sum_{j \neq i}\left(E_{ij}^{\text{REBO}} + E_{ij}^{\text{MORSE}} + \sum_{k \neq i,j}\sum_{l \neq i,j,k} E_{ijkl}^{\text{TORSION}}\right) \tag{2.10}$$

where

E_{ij}^{REBO} REBO term
E_{ij}^{MORSE} MORSE term
$E_{ijkl}^{\text{TORSION}}$ TORSIONAL energy term

and the Morse term is modelled as

$$U_{ij}(r) = -\varepsilon_{ij}\left[1 - \left(1 - e^{-\alpha_{ij}\left(r - r_{ij}^{eq}\right)}\right)^2\right] \tag{2.11}$$

The parameters ε and r^{eq} represent the minimum energy values (depth and location), while α refers to the adjusted curvature of the potential energy at the point of minimum separation [6]. The interlayer distance between the two graphene layers was maintained at 3.4 Å, with the separation modelled using the Lennard–Jones potential. The energy parameter was set to 0.00239 eV, and the distance parameter was 2.5 Å [9].

2.1.3 EAM/Alloy Interatomic Potential

The AlCoCrFeNi HEA [10] [15] discussed in Chaps. 7 and 8 are modelled by utilizing EAM/alloy interatomic potential [11]. Embedded-atom method (EAM) force field is capable of modelling the pairwise interactions for metals and metal alloys. The total energy E_i of a constituent atom i is modelled as

$$E_i = F_\alpha\left(\sum_{j\neq i}\rho_\beta\left(r_{ij}\right)\right) + \frac{1}{2}\sum_{j\neq i}\phi_{\alpha\beta}\left(r_{ij}\right) \tag{2.12}$$

where F denotes the embedding energy which is a function of the atomic electron density r, φ is a pair potential interaction, and a and b are the element types of i and j. The gold (Au) nano-projectile (used in *Chap.* 9) is modelled with an *EAM* force field. The interface between constituent elements of HEA and the projectile (Au) is modelled by using the *Lennard–Jones* (*L–J*) force field; the calculation of the *L–J* parameters is done as shown by Tang and Li [12]. The *LJ* parameters utilized in the simulations are presented in Table 2.1.

Table 2.1 *L–J* parameters used to model the interaction between HEA configurations and Au nano-projectile [15]

Atom pair	ε (eV)	σ (Å)
Al–Au	70.62	3.47
Co–Au	11.76	2.74
Cr–Au	12.17	2.81
Fe–Au	11.33	2.76
Ni–Au	12.17	2.72

2.1.4 Temperature and Pressure Control

As in the lab experiments, some constraints are imposed on the system to determine the desired properties. Similarly, in the simulation, it is required to adjust the state of the simulation by controlling temperature and pressure. During MD simulations, temperature control is performed by systematically manipulating the kinetic energy of the system, specifically by scaling the atomic velocities using different thermostat algorithms such as Anderson, Berendson, and Nose–Hoover. Thermostats follow the given relation [13]:

$$T = \frac{1}{3Nk_{\mathrm{B}}} \sum_i m_i v_i^2 \tag{2.13}$$

where k_{B} and T refer to the Boltzmann constant and the instantaneous temperature of the system, respectively. As the velocity fluctuations are inevitable in the MD simulation, the thermostat algorithm maintains the average temperature of the system constant.

Pressure control is imparted by using the generalized equipartition theorem and is defined as [13]:

$$P = \frac{1}{V}\left[Nk_{\mathrm{B}}T + \frac{1}{3}\left\langle \sum_i r_i.F_i \right\rangle\right] \tag{2.14}$$

Similar to the thermostat algorithms, the barostat algorithms such as the Berendsen or the Nose–Hoover barostat also keep the average value of pressure constant. The pressure control is achieved by keeping the volume of the system in check.

2.1.5 Statistical Ensembles

Statistical ensembles refer to all the possible thermodynamic states of the system. The state of the system is defined by the set of observable thermodynamic quantities such as temperature and pressure, and these observable quantities are in equilibrium.

The microcanonical ensemble (NVE) is the most commonly used in MD and keeps the number of atoms, volume, and energy of the system constant. To impart constant pressure and constant temperature to the system, isothermal–isobaric ensemble (NPT) is used where P and T stand for constant pressure and temperature. The canonical ensemble (NVT) is used to simulate systems under fixed volume and temperature. Figure 2.2 graphically demonstrates the characteristics of the NVE, NVT, and NPT ensemble [14].

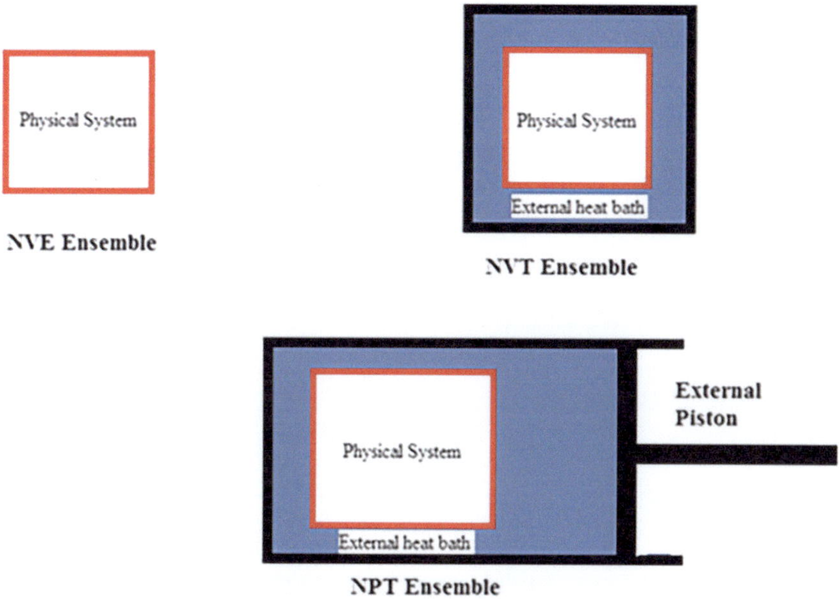

Fig. 2.2 Physical significance of NVE, NVT, and NPT ensembles

2.2 Designing Nanoscale Materials

As mentioned earlier, material systems within molecular dynamics (MD) simulation frameworks are modelled as collections of individual atoms. In the case of graphene, these atoms are arranged in a specific pattern: a 2D, single-atom-thick layer of carbon atoms organized in a hexagonal lattice. In contrast, for crystalline high-dimensional materials such as high-entropy alloys (HEAs), atoms are distributed within a defined simulation domain based on specified dimensions. The concentration of individual elements, lattice constant, and initial crystal structure are controlled within the input script of the simulation. Generally, the modelled material's atomic configurations are represented as atom types (individual elements), and atomic coordinates (x, y, and z) of the constituent elements.

2.2.1 Modelling Graphene in LAMMPS

MD simulation in LAMMPS requires execution of certain tasks by using the perfectly written input script. The input script for modelling the graphene within LAMMPS is illustrated in Fig. 2.3.

```
# Generate 10 nm x 10 nm graphene and save to graphene.dat
units           metal
dimension       3
boundary        p p p
atom_style      atomic

# Graphene lattice: hexagonal with lattice constant ~2.46 Å
lattice         hex 2.46 orient x 1 0 0 orient y 0 1 0 orient z 0 0 1

# Convert 100 nm to Å: 10 nm = 100 Å
region          graphene_region block 0 100 0 100 0 10 units box

# Create box and atoms
create_box      1 graphene_region
create_atoms    1 region graphene_region

# Define AIREBO potential
pair_style      airebo 3.0
pair_coeff      * * CH.airebo C

# Write the structure to file
write_data      graphene.dat
```

Fig. 2.3 LAMMPS input script to design squared-shaped graphene layer (10 nm × 10 nm)

The input script shown in Fig. 2.3 uses metal *units*, which ensure the length units in angstroms (Å), energy units in eV, time units in picoseconds (ps), and temperature units in Kelvin (K). Further, the input script defines the dimension as 3, periodic boundary conditions along the x, y and z axis, and atomic configuration of the graphene model. The graphene layer is modelled by using the lattice constant of 2.46 (relevant to graphene), and hex lattice structure. The carbon atoms are arranged in a hexagonal fashion, within the defined region of 0 to 100 Å along X axis, and 0 to 100 Å along Y axis. The input script utilizes the AIREBO force field to assign the atom type of carbon. Finally, the modelled graphene sheet is saved into LAMMPS usable format as *graphene.dat*.

2.2.2 Modelling AlCoCrFeNi HEA in LAMMPS

The LAMMPS input script (refer to Fig. 2.4) models an equiatomic AlCoCrFeNi high-entropy alloy (HEA) using an FCC lattice with a lattice constant of 3.56 Å. The simulation box spans 60 lattice units along the X, Y, and Z axes, creating a sufficiently large atomic system. Interatomic interactions are defined using the EAM/alloy potential, which is well-suited for metallic systems like HEAs. To achieve equiatomic composition, atomic types are assigned in successive steps using the set type/fraction command, ensuring each element is distributed according to its target concentration. As an outcome, the atomic configurations of the modelled HEA are saved in the LAMMPS usable file AlCoCrFeNi_HEA.dat.

```
# LAMMPS input script to generate AlCoCrFeNi HEA structure and write data file
units           metal
dimension       3
boundary        p p p
atom_style      atomic

# Simulation box dimensions in lattice units
variable        x_length equal 60
variable        y_length equal 60
variable        z_length equal 60

# FCC lattice constant (typical for AlCoCrFeNi HEAs)
variable        lattice equal 3.56

# Element composition (equiatomic)
variable        fFe equal 20
variable        fNi equal 20
variable        fCr equal 20
variable        fCo equal 20
variable        fAl equal 20

# Create FCC lattice and simulation region
lattice         fcc ${lattice}
region          simbox block 0 ${x_length} 0 ${y_length} 0 ${z_length} units lattice
create_box      5 simbox
create_atoms    1 region simbox

# Randomly assign atom types based on fractions
set type 1 type/fraction 2 $((v_fNi+v_fCr+v_fCo+v_fAl)/100.0)       12345
set type 2 type/fraction 3 $((v_fCr+v_fCo+v_fAl)/(v_fNi+v_fCr+v_fCo+v_fAl)) 12346
set type 3 type/fraction 4 $((v_fCo+v_fAl)/(v_fCr+v_fCo+v_fAl))      12347
set type 4 type/fraction 5 $((v_fAl)/(v_fCo+v_fAl))                   12348

# Assign masses to each element
mass 1 55.845   # Fe
mass 2 58.6934  # Ni
mass 3 51.9961  # Cr
mass 4 58.9332  # Co
mass 5 26.9815  # Al

# Define potential (EAM alloy format)
pair_style      eam/alloy
pair_coeff      * * FeNiCrCoAl.eam.alloy Fe Ni Cr Co Al

# Neighbor settings
neighbor        2.0 bin
neigh_modify    delay 5

# Save structure to data file
write_data      AlCoCrFeNi_HEA.dat
```

Fig. 2.4 LAMMPS input script to model equiatomic AlCoCrFeNi HEA

2.3 LAMMPS Input Scripts for Performing Atomistic Simulations

The '*input script*' is a key component for executing the LAMMPS simulation. All the instructions that a program must execute are scripted in the input script. The input script is structured in the following manner:

- Initialization
- Atom and Lattice definition

- Force fields
- Settings
- Run.

The simulation process begins by defining fundamental settings such as the unit system, the dimensionality of the simulation (2D or 3D), and the boundary conditions (periodic or non-periodic). Next, the lattice parameters are specified, including the lattice type and lattice constant, which determine the crystal structure and spacing of atoms. Based on these parameters, the size of the simulation box is set, and atoms are initialized within the defined region.

Following the structural setup, an appropriate interatomic potential (force field) is chosen to model how atoms interact with one another. Once the atomic structure and interaction model are defined, simulation controls are established. These include selecting the physical quantities to be monitored during the simulation—such as potential energy, temperature, pressure, and cohesive energy. Finally, a time step is set, and the simulation is executed for a sufficient number of steps to capture the desired physical behaviour. The detailed procedures for performing MD simulations on the specific nanomaterials explored in this book are presented in Chaps. 5 through 9.

2.4 Summary

This chapter focuses on the foundational aspects of conducting molecular dynamics (MD) simulations for the nanomaterials explored throughout the book. It begins by establishing the essential theoretical preliminaries of MD, followed by a detailed understanding of the critical components such as interatomic potentials, temperature and pressure control, and statistical ensembles. In addition to that, the chapter also discusses the structure of the LAMMPS input script, especially used to model graphene and high-entropy alloys in an MD environment. In summary, this chapter outlines the key considerations for performing MD simulations on two distinct classes of nanomaterials: graphene and high-entropy alloys (note that these are discussed in the following chapters as representative low- and high-dimensional materials).

References

1. Lee JG (2016) Computational materials science: an introduction. CRC Press
2. Gupta KK, Roy A, Dey S (2020) Comparative study of various defects in monolayer graphene using molecular dynamics simulation. In: Advances in applied mechanical engineering: select proceedings of ICAMER 2019. Springer Singapore, pp 539–546
3. Roy A, Gupta KK, Dey S (2022) Probabilistic investigation of temperature-dependent vibrational behavior of hetero-nanotubes. Appl Nanosci 12(7):2077–2089

4. Gupta KK, Barman S, Dey S, Naskar S, Mukhopadhyay T (2024) On exploiting nonparametric kernel-based probabilistic machine learning over the large compositional space of high entropy alloys for optimal nanoscale ballistics. Sci Rep 14(1):16795
5. Tersoff JJPRB (1989) Modelling solid-state chemistry: interatomic potentials for multicomponent systems. Phys Rev B 39(8):5566
6. O'Connor TC, Andzelm J, Robbins MO (2015) AIREBO-M: a reactive model for hydrocarbons at extreme pressures. J Chem Phys 142(2):024903
7. Shoaib H, Peng Q, Alsayoud AQ (2021) Atomic insights into fracture characteristics of twisted tri-layer graphene. Crystals 11(10):1202
8. Haque BZG, Chowdhury SC, Gillespie JW Jr (2016) Molecular simulations of stress wave propagation and perforation of graphene sheets under transverse impact. Carbon 102:126–140
9. Zhang X, Chen Z, Chen H, Xu L (2021) Comparative studies of thermal conductivity for bilayer graphene with different potential functions in molecular dynamic simulations. Results in Physics 22:103894
10. Gupta KK, Barman S, Dey S, Mukhopadhyay T (2024) Explainable machine learning assisted molecular-level insights for enhanced specific stiffness exploiting the large compositional space of AlCoCrFeNi high entropy alloys. Mach Learn: Sci Technol 5(2):025082
11. Farkas D, Caro A (2020) Model interatomic potentials for Fe–Ni–Cr–Co–Al high-entropy alloys. J Mater Res 35(22):3031–3040
12. Tang Y, Li DY (2022) Dynamic response of high-entropy alloys to ballistic impact. Sci Adv 8(31):eabp9096
13. Mawyin T (2015) Sensitivity analysis of material property predictions in molecular dynamics with interatomic potentials under uncertainty. University of Toronto (Canada); Ray JR, Rahman A (1984) Statistical ensembles and molecular dynamics studies of anisotropic solids. J Chem Phys 80(9):4423–4428
14. Gupta KK, Dey S (2019) Effect of temperature on the fracture strength of perfect and defective monolayered graphene. In: Advances in computational methods in manufacturing: select papers from ICCMM 2019. Springer Singapore, pp 793–804
15. Gupta KK, Barman S, Dey S, Naskar S, Mukhopadhyay T (2024) On exploiting nonparametric kernel-based probabilistic machine learning over the large compositional space of high entropy alloys for optimal nanoscale ballistics. Sci Rep 14(1):16795

Chapter 3
Machine Learning

3.1 Overview

This chapter outlines the operational principles of the machine learning (ML) algorithms utilized in the subsequent chapters of this book. It emphasizes how computationally efficient ML models are integrated with insights from extensive molecular dynamics (MD) simulations to achieve a deeper understanding of material behaviour, while simultaneously reducing the computational costs typically associated with large-scale MD studies. A detailed conceptual overview is provided for some of the technologically advanced ML models, including ensemble learning techniques (such as bagging and boosting) and Gaussian Process Regression (GPR). The chapter also introduces a standardized ML pipeline tailored for ML-driven atomistic design of materials.

Machine Learning (ML) is a subset of artificial intelligence (AI) which provides an efficient platform to develop surrogate models for predicting physical responses by significantly reducing the need to perform large-scale experiments/simulations. By utilizing an appropriate ML model, the linear/non-linear relationship between decision variables and the desired target response is established by executing a series of data transformation tasks. The evolution of computational technologies and the simultaneous emergence of knowledge representation methods (in mathematical forms) has grown the interest of research communities across various domains to utilize the ML model-driven predictive frameworks for initial screening of the solutions. The technological advances, in the past few years, have enabled the materials research community to develop ML-driven frameworks for accelerated design of functional materials [1–6]. In this pursuit, the evolution of ML-driven predictions has demonstrated great potential in exploring the unchartered territory of materials.

Machine learning (ML) algorithms can be primarily categorized into *supervised ML, unsupervised ML, and reinforcement ML* approaches (refer to Fig. 3.1). The supervised machine learning (ML) approach relies on prior knowledge obtained from properly labelled data or recorded system responses to establish relationships

K. K. Gupta et al., *Machine Learning in Nanoscale Materials Design*, Materials Horizons: From Nature to Nanomaterials, https://doi.org/10.1007/978-981-95-2660-4_3

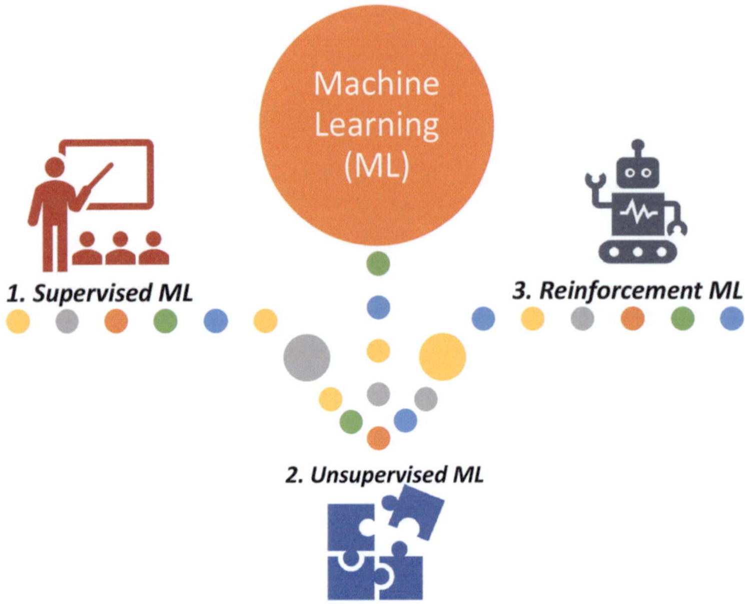

Fig. 3.1 Broad categorization of Machine Learning models

between input features and desired outputs. In contrast, unsupervised ML aims to uncover patterns within unstructured data using techniques such as clustering and dimensionality reduction. Finally, reinforcement learning involves training models through a system of rewards and penalties—commonly used in robotics—to enable the model to perform the desired task within a specific scenario.

The investigations carried out in this book mostly utilize the concepts of supervised ML. In the cases of supervised ML, the aim is to establish the relationship between dependent and independent variables with the help of labelled classes (in classification problems) or the continuous values of the dependent variables (in regression problems). In general, the supervised ML approach is guided by the available information in terms of input and target variables. Such ML models play a vital role in the development of the predictive framework, which is capable of predicting the target values for any random configuration of the independent variables. The supervised ML models can be further classified as follows [7].

3.1.1 Parametric ML Models

The ML models which provide a closed-form solution as a mathematical equation are known as parametric models. Such models result in a mathematical relationship between the target variable (TV) and input features (IF: IF_1, IF_2, IF_3, IF_4, etc.) as

$$\text{TV} = \beta_0 + \text{IF}_1\beta_1 + \text{IF}_2\beta_2 + \text{IF}_3\beta_3 + \text{IF}_4\beta_4 + \cdots \quad (3.1)$$

where $\beta_{(0\text{–}4\ldots)}$ values denotes the coefficients. It is a general practice to use parametric ML models in cases where the correlation between the target variable and input features is known. Basic models such as simple linear regression [8], multivariate linear regression [8], polynomial regression [9], and logistic regression [10] (for classification tasks) come under the category of parametric ML models. The parametric models are computationally efficient and offer ease in interpretability, but these models mostly lack in capturing the complex relationships between the input features and the target response.

3.1.2 Non-parametric ML Models

Unlike parametric ML models, the non-parametric ML models do not offer the closed-form solution (mathematical equation) of the relationship between the input features and the target response; rather these models utilize probabilistic algorithms to efficiently capture the trend of the dataset. In terms of materials informatics, these models demonstrate a promising potential in capturing the unknown intricate physical phenomena of the materials. For example, a parametric ML model establishes the relationship between the alloy composition and wear rate by assigning specific weights to each compositional element, reflecting their importance. In contrast, a non-parametric model provides a probabilistic distribution of potential solutions rather than a single, fixed outcome. This flexibility makes non-parametric models more adaptable to changes in the data [11]. Some of the examples of non-parametric models are K-nearest neighbour (KNN) [12], support vector machine (SVM) [13], decision trees [14], Gaussian process regression (GPR) [15], etc. In the example of alloy composition-dependent wear rate, the plan of attack that KNN follows is that it finds the solution by averaging for the 'k' number of the neighbours in the vicinity of the available training data, whereas decision trees (DT) work by recursively splitting the training data into subsets by using the levels of the different features involved (alloy composition) and perform the averaging of the solutions to predict for the unseen data. The SVM employs the kernel trick to represent the dataset in higher dimensional space for obtaining the flat hyperplane with the minimum errors possible, which makes the SVM model less prone to overfitting. The GPR model utilizes the Bayesian approach by establishing the relationship between input features and target responses by providing the probability distribution over functions, which facilitates the evaluation of the uncertainty associated with the model predictions. Figure 3.2 provides illustrations of the aforementioned ML models.

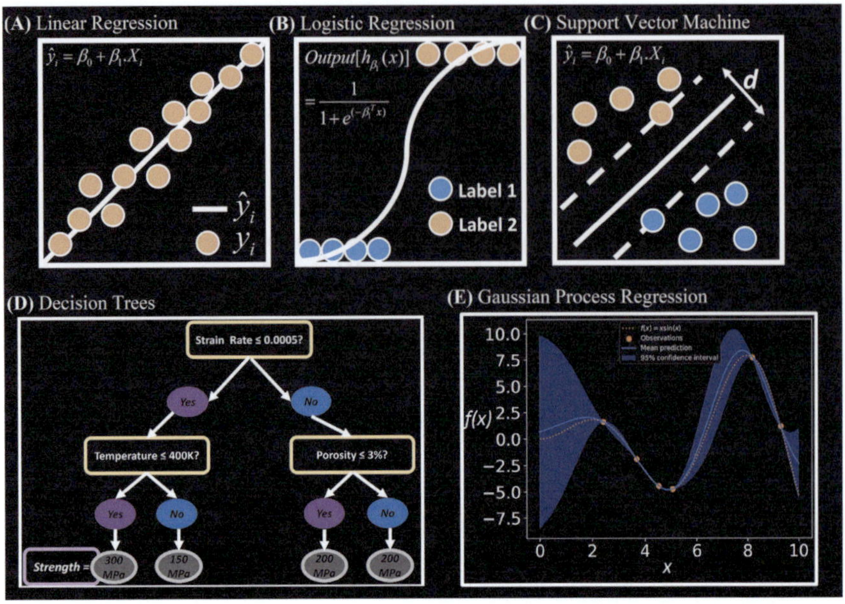

Fig. 3.2 Different machine learning algorithms: **a**, **b** parametric machine learning; **c–e** non-parametric ML models

3.2 Machine Learning Models

Advances in computational capabilities have enabled the materials research community to adopt sophisticated learning algorithms for materials characterization, such as artificial neural networks (ANNs) [3, 16], graph neural networks (GNNs) [17], and convolutional neural networks (CNNs) [18]. While these models excel at simulating human-like learning, they often require large, high-quality datasets. In contrast, advanced machine learning (ML) frameworks, such as ensemble methods, Bayesian inference, and Gaussian processes, can effectively capture material-specific trends even when data is sparse or of low quality. Building on this insight, this book emphasizes the implementation of active learning frameworks grounded in these robust ML models. As outlined in the preceding section, parametric machine learning (ML) models offer advantages in terms of simplicity and interpretability. However, they often fall short of capturing complex relationships between input features and target responses. These limitations can be effectively addressed through the use of non-parametric ML models. With this understanding, this book employs several non-parametric ML approaches to uncover data-driven insights for the comprehensive characterization of graphene and high-entropy alloys. This section provides a foundational understanding of various machine learning (ML) algorithms that are applied throughout the subsequent chapters of this book. Broadly, computationally efficient ML models have been developed and integrated with insights derived from extensive molecular dynamics (MD) simulations. This integration aims to uncover a deeper understanding of material behaviour while also reducing the high computational cost typically associated with large-scale MD simulations.

Chapters 5, 6, and 9 focus on the utilization of ensemble ML models, while Chaps. 7 and 8 demonstrate the successful implementation of the Gaussian Process Regression (GPR) model. This highlights the successful utilization of non-parametric ML models, such as different variants of ensemble ML models and GPR, across the subsequent chapters. This choice is justified by the highly non-linear relationships present between the input and output configurations in the studied materials systems. Therefore, the following subsections provide a detailed mathematical and algorithmic overview of ensemble machine learning models and Gaussian Process Regression (GPR).

3.2.1 Ensemble Machine Learning

Ensemble ML algorithms are a special class of models which utilizes several base models also known as '*weak models*', or '*weak learners*' to gradually increase the predictive performance of the final model [19]. These '*weak learners*' can be decision trees (DT). In Decision Trees (DT), instances with similar attributes are grouped during the training process. Once the model is built, new or unseen samples are assigned to one of these predefined clusters based on the similarity of their features. In regression tasks, a similar strategy is applied, where the input space is recursively divided into disjoint regions using binary partitioning (refer to Fig. 3.3). This enables the model to make predictions based on the average output value within each region [14].

The functional behaviour of a DT regressor is effectively illustrated in Fig. 3.3. In this figure, X_1 and X_2 represent two independent input parameters, while the response variable Y is modelled as a function of both. The observed values of Y, corresponding to various combinations of X_1 and X_2, are shown as scattered points. Based on these observations, a DT performs a series of data splits to cluster similar instances. For example, **Split-1** is created by applying a condition such as $\boldsymbol{X_1 < 33}$ and $\boldsymbol{X_2 < 42}$. Similarly, **Split-2** uses conditions like $\boldsymbol{33 < X_1 < 41}$ and $\boldsymbol{0 < X_2 < 100}$. This process of recursive partitioning continues, dividing the input space into distinct, non-overlapping regions. Each split forms a branch, and the data points within each region are referred to as leaves. In the subsequent step, the average of the target values (Y) within each leaf is computed, providing a representative output for that region. This structure forms the basis of a decision tree, where the trained model predicts the response for new, unseen data by identifying the appropriate branch and using the corresponding average value of that leaf. In ensemble machine learning models, multiple such decision trees act as weak learners, each capturing local patterns in the data. Together, they contribute to a stronger, more generalized predictive model. The underlying idea is that a group of models working together can achieve better generalization more efficiently than any individual model on its own.

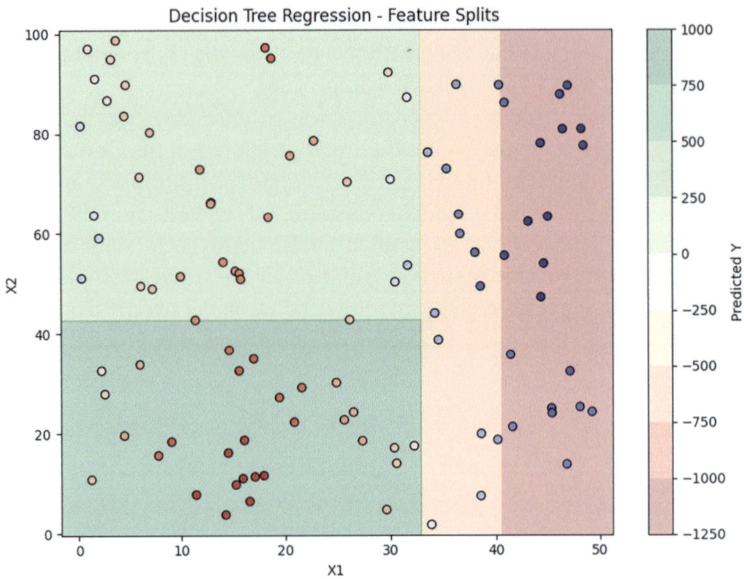

Fig. 3.3 Splitting of the decision nodes during Decision Tree formation

Ensemble methods are particularly useful for solving complex tasks where single models struggle to capture accurate mathematical representations of the data. The ensemble ML models are categorized into three different approaches as follows.

3.2.1.1 Bagging Regressor

The term Bagging is a made-up word by combining Bootstrap AGGregatING; as the name suggests, bagging regressors utilize two different approaches, namely bootstrapping and aggregation [20]. The pseudocode for performing bagging operation is illustrated in Fig. 3.4. The bagging mechanism starts with the bootstrapping operation, wherein the random subsets of the training dataset are iteratively selected to train B numbers of base regressors (such as decision trees). Individual DT regressors (M_1 to M_B) are developed and utilized for the aggregation stage. In the aggregation phase of ensemble learning, the final prediction is obtained by averaging the outputs of all base learners, thereby producing a consolidated and more robust result. Bagging-based ensemble models provide substantial benefits, including variance reduction and improved resistance to overfitting. Additionally, their ability to train base learners in parallel facilitates quicker convergence and generalization.

```
# Pseudo-code for Bagging Regressor
Input:
  - Training data X = {x1, x2, ..., xn}
  - Target values Y = {y1, y2, ..., yn}
  - Number of base regressors: B
  - Base Regressor: base_model (e.g., Decision Tree Regressor)

Training Phase:
  For i = 1 to B:
    1. Create a bootstrap sample (X_i, Y_i) by sampling with replacement from (X, Y)
    2. Train a base regressor M_i on (X_i, Y_i)
    3. Store M_i in the ensemble

Prediction Phase:
  For a new input x:
    1. Predict ŷ_i = M_i(x) for all i = 1 to B
    2. Return final prediction: ŷ = (1/B) * Σ ŷ_i
```

Fig. 3.4 Pseudocode for execution of the bagging operation of ensemble Machine Learning

3.2.1.2 Boosting Regressor

In the context of ensemble machine learning, boosting is another powerful technique particularly effective at reducing bias during model training. Similar to bagging regressors, boosting regressors are also composed of multiple base models (weak learners). However, unlike bagging—which trains weak learners in parallel—boosting trains them sequentially. Each subsequent model is trained by placing greater emphasis on the errors made by its predecessor, thereby progressively improving overall predictive performance [21]. The underlying idea is to focus more on data points that were previously predicted poorly.

The boosting regressor utilizes the iterative loop of perfecting the weak-learner's predictions for T number of rounds; during each iteration, pseudo residuals (difference between actual response y_i (for ith data point) and prediction of current ensemble model $F_{(t-1)}$) are evaluated (refer the pseudocode provided in Fig. 3.5). These residuals are fed to the next stage of gradient decent optimization for minimizing the loss function $h_t(x)$ to zero by using the equation:

$$F_t(x) = F_{t-1}(x) + \eta . h_t(x) \tag{3.2}$$

where η stands for learning rate. With this approach, the gradient $h_t(x)$ is sequentially minimized towards zero, allowing each new model to improve upon the predictive performance of the previous one. Light Gradient Boosting Machine (LGBM) and Extreme Gradient Boosting (XGBoost), both the ML models, utilize the same boosting idea, with distinct differences in tree building strategy, LGBM utilizes leaf-wise tree building within stipulated depth, and XGBoost utilizes level-based tree building. Both the models consider the L_1 and L_2 regularization techniques to prevent the model from overfitting. The objective function of boosting regressor looks like [22]

```
# Pseudo-code for Boosting Regressor
Input:
  - Training data: X = {x1, x2, ..., xn}
  - Target values: Y = {y1, y2, ..., yn}
  - Number of boosting rounds: T
  - Base learner: WeakRegressor (e.g., decision tree)
  - Learning rate: η

Initialize:
  - Start with initial prediction: F_0(x) = mean(Y)

For t = 1 to T:
  1. Compute pseudo-residuals:
     rᵢ = yᵢ - F_{t-1}(xᵢ) for all i

  2. Train base model h_t(x) on (X, r) — fit residuals

  3. Update the ensemble prediction:
     F_t(x) = F_{t-1}(x) + η * h_t(x)

Final Prediction:
  Return F_T(x) for any input x
```

Fig. 3.5 Pseudocode for boosting operation performed in ensemble Machine Learning model

$$\mathrm{Obj} = \sum_{i=1}^{n} l(y_i, \hat{y}_i) + \sum_{k=1}^{N_{\mathrm{tree}}} \Omega(f_k) \tag{3.3}$$

where $l(y_i, \hat{y}_i)$ is a differentiable loss function evaluating the residuals (difference of actual and predicted response), $\Omega(f_k)$ is a regularization term that penalizes model complexity, n refers to the number of training samples, and N_{tree} denotes the number of trees. The objective function consists of two components: the loss function $\sum_{i=1}^{n} l(y_i, \hat{y}_i)$ and the regularization term $\sum_{k=1}^{N_{\mathrm{tree}}} \Omega(f_k)$.

The regularization term can be explained as

$$\Omega(f_k) = \gamma T + \frac{1}{2}\lambda \sum_{j=1}^{T} w_j^2 \tag{3.4}$$

where T is the number of leaves in the tree, w_j is the leaf weights, γ is a regularization parameter controlling tree complexity, and λ is the L_2 regularization coefficient.

3.2.2 *Gaussian Process Regression (GPR)*

The kernel-based model such as Gaussian Process Regression (GPR) makes use of non-parametric approach to map the variation in input features with the desired response. Instead of providing a unique prediction, GPR aims to produce the uncertainty bounds associated with the prediction, which makes this model more reliable than others. The primary objective of this model is to map the desired target variable to a regression function, incorporating normally distributed noise. For instance, $X = x_1, x_2, x_3, \ldots, x_n$ represents the input feature space, and $Y = y_1, y_2, y_3, \ldots, y_n$ represents the target variables. The GPR assumes that a GP with a mean function (μ) and covariance function (k) produces a function f which is capable in successfully connecting X with Y. Such function f produces a distribution for a set of test data (x^*) as follows [23]:

$$f(x*) \sim N(\mu(x*), k(x*, x*)) \tag{3.5}$$

The kernel function utilizes the covariance function to highlight the extent of uncertainty in the predictions, for example, the commonly used squared exponential kernel, which is defined as [24]

$$k\left(x_i, x_j\right) = \sigma^2 \exp\left(-\frac{\left\|x_i - x_j\right\|^2}{2l^2}\right) \tag{3.6}$$

Here, $k(x_i, x_j)$ is a *kernel function* which evaluates the similarity between two different data points (x_i and x_j); σ^2 defines the kernel's variance: higher value results in higher vertical spread; it also regulates how strongly the data points correlate with each other. $\left\|x_i - x_j\right\|^2$ is the squared Euclidean distance between the data points, $x_i = x_j$ (closely correlated data points) will yield this value 0, and $\exp\left(-\frac{\left\|x_i - x_j\right\|^2}{2l^2}\right)$ will be 1; likewise loosely correlated data points x_i and x_j will yield $\exp\left(-\frac{\left\|x_i - x_j\right\|^2}{2l^2}\right)$ close to zero. Here, l is the length scale parameter of the kernel which controls how quickly the similarity between the data points decay with the distance. A lower value of l makes the kernel decay faster.

The GPR utilizes the Bayesian inference to produce the distribution of the function f which is highly likely to produce the given dataset (X, Y). The posterior distribution of f is given by

$$\begin{aligned} p(f|X, Y) &= \frac{p(Y|X, f)p(f)}{p(Y|X)} \\ p(Y|X)&: \quad \text{marginal probability} \\ \text{here } p(f)&: \quad \text{prior distribution} \\ p(Y|X, f)&: \text{likelihood} \end{aligned} \tag{3.7}$$

By learning the posterior distribution of f, the model evaluates the posterior predictive distribution over the unknown data (test data, X^*), which is defined as

$$p(f * |X*, Y, X) = \int p(f * |X*, f), p(f|Y, X)\mathrm{d}f \tag{3.8}$$

$p(f * |X*, Y, X)$ provides the conditional probability of the distribution of predicted function f^*, given the training data (X, Y) and set of test data (X^*).

3.3 A Generalized Machine Learning Pipeline

This section discusses the generalized ML pipeline incorporated in the subsequent chapters (Chaps. 5–9) of this book. The pipeline consists of the sequential steps required to train and assess the predictive performance of the ML models. Given the computationally expensive and intensive nature of performing large-scale MD simulations, it is critical to developing such frameworks that not only accelerate material discovery but also reveal deep insights into material behaviour that would otherwise remain unexplored. The ML-driven investigation always starts with the appropriate sample generation scheme, which will efficiently capture the complete continuous domain of the parametric range of variation while conducting the atomistic simulations. In this regard, at first, the randomly distributed samples are generated within the parametric range of variation. The mathematical formulation of the random sampling to construct the sample space can be represented as follows [3, 25]:

$$p_i = p_{\min} + r_i(p_{\max} - p_{\min}) \tag{3.9}$$

where $(p_{\max} - p_{\min})$ denotes the range of variation of the considered input parameter (p), $p_{\min}$ represents the minimum value of the input parameter, and r_i denotes the random number drawn from the probabilistic distribution in the range of 0 to 1. The variable p_i provides the ith realization of the sampled parameter, where $i \in (1{:}N)$, N denoting the total number of samples considered. The sample space generated by random sampling is fed into the MD simulation environment to obtain the desired responses. The constructed dataset is utilized to train and validate the machine learning models. The framework utilized to integrate molecular dynamics simulation with the machine learning algorithms is illustrated in Fig. 3.6 by means of a flowchart.

The steps involved during the ML model development include normalization of the input features, by using the z-score, given by

$$X_{i,\text{norm}} = \frac{X_i - \mu(X)}{\sigma(X)} \tag{3.10}$$

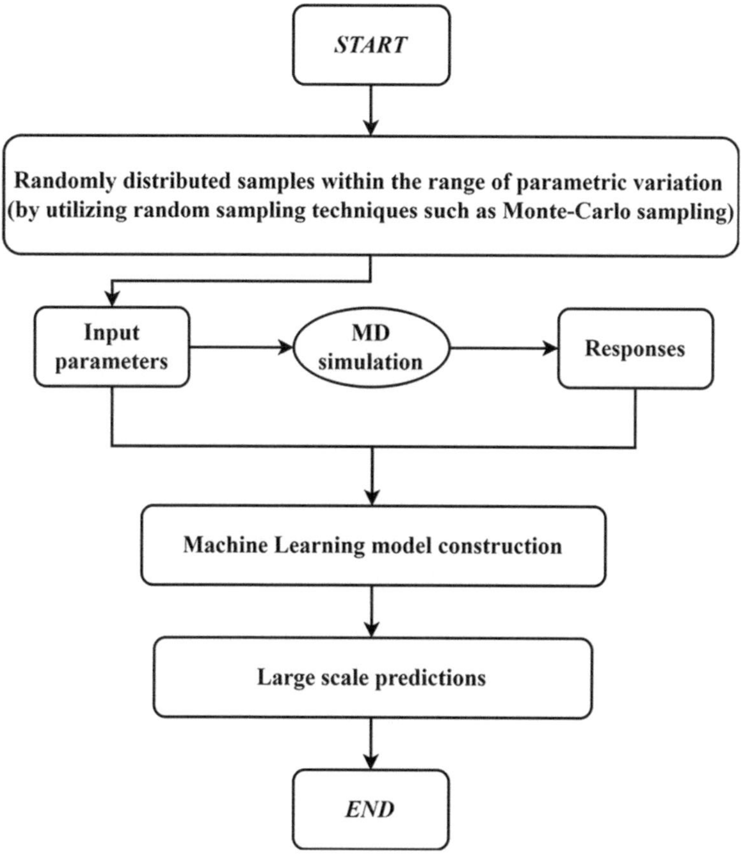

Fig. 3.6 Flowchart of Machine Learning-assisted molecular dynamics simulation

where $\mu(X)$ denotes the mean of the input feature X and $\sigma(X)$ denotes the standard deviation of the input feature X. Normalization of input features is a crucial preprocessing step to mitigate differences in scale among the input variables. The normalized feature matrix $[X]$, along with the corresponding target vector $[y]$, is then randomly split into training and testing datasets using an appropriate ratio (e.g. 90:10, 80:20). The training data is subsequently used to develop the model through the application of k-fold cross-validation. In this technique, the training dataset is divided into k equally sized folds. During each iteration, one fold is set aside for validation, while the remaining $(k-1)$ folds are used to train the model. This process is repeated k times, ensuring that each fold is used once as a validation set. This cross-validation strategy enables a simultaneous training–validation mechanism, promoting improved generalization and reducing the risk of overfitting. Furthermore, the training process includes an iterative loop for hyperparameter tuning, where multiple combinations of hyperparameters are evaluated. The optimal set is selected based on the configuration that yields the lowest prediction error across all k-folds, resulting in a well-generalized and performance-optimized machine learning model. The hyperparameter-tuned ML model is further tested for accuracy by utilizing the test data (unseen data). The accuracy obtained on the out-of-fold samples—those not used during the training phase—reflects the model's true predictive capability and indicates its susceptibility to making incorrect predictions.

The generalization capability of the constructed model is assessed on the basis of the values of correlation coefficient (R^2) and relative error (%). The correlation coefficient is defined as [1]

$$R^2 = 1 - \frac{\sum_{i=1}^{n} (y_i - f_i)^2}{\sum_{i=1}^{n} (y_i - \overline{y})^2} \tag{3.11}$$

where y_i, f_i, and $\bar{y}_i$ refer to the observed value, predicted value, and mean of the observed value, respectively, and i denotes the sample index. The value of R^2 provides the statistical signification of the closeness of the predicted points to the regression line. It determines the goodness of fit for any model, and its value is always between 0 and 1. The percentage error in the predicted responses is evaluated as

$$\text{error}(\%) = \left[\frac{y_{i,\text{MD simulation}} - y_{i,\text{ML}}}{y_{i,\text{MD simulation}}} \right] \times 100 \tag{3.12}$$

where $y_{i,\text{ MD simulation}}$ denotes the true response derived from MD simulation, and $y_{i,\text{ ML}}$ denotes the predicted response derived from machine learning.

3.4 Summary

This chapter establishes the foundational understanding of the machine learning (ML) models used throughout the subsequent chapters of this book. It also introduces the integrative framework for developing ML models driven by molecular dynamics (MD) simulations. The chapter begins by categorizing ML models based on their mathematical representation. It then delves into non-parametric models, with a detailed explanation of ensemble ML models and the Gaussian Process Regression (GPR) model. Finally, the chapter concludes with a discussion on a general ML pipeline that can be used in conjunction with MD simulations for the ML-driven design of materials.

References

1. Gupta KK, Mukhopadhyay T, Roy A, Roy L, Dey S (2021) Sparse machine learning assisted deep computational insights on the mechanical properties of graphene with intrinsic defects and doping. J Phys Chem Solids 155:110111
2. Gupta KK, Mukhopadhyay T, Roy L, Dey S (2022) High-velocity ballistics of twisted bilayer graphene under stochastic disorder. Adv Nano Res 12(5):529–547
3. Roy A, Gupta KK, Dey S (2022) Probabilistic investigation of temperature-dependent vibrational behavior of hetero-nanotubes. Appl Nanosci 12(7):2077–2089
4. Gupta KK, Roy L, Dey S (2022) Hybrid machine-learning-assisted stochastic nano-indentation behaviour of twisted bilayer graphene. J Phys Chem Solids 167:110711
5. Gupta KK, Roy A, Mukhopadhyay T, Roy L, Dey S (2022) Probing the stochastic fracture behavior of twisted bilayer graphene: efficient ANN based molecular dynamics simulations for complete probabilistic characterization. Mater Today Commun 32:103932
6. Dey S, Mukhopadhyay T, Adhikari S (2018) Uncertainty quantification in laminated composites: a meta-model based approach. CRC Press
7. Imam F, Musilek P, Reformat MZ (2024) Parametric and nonparametric machine learning techniques for increasing power system reliability: a review. Information 15(1):37
8. Su X, Yan X, Tsai CL (2012) Linear regression. Wiley Interdiscip Rev: Comput Statis 4(3):275–294
9. Ostertagová E (2012) Modelling using polynomial regression. Procedia Eng 48:500–506
10. LaValley MP (2008) Logistic regression. Circulation 117(18):2395–2399
11. Krishnan NA (2024) Machine Learning for materials discovery: numerical recipes and practical applications. Springer Nature
12. Samet H (2007) K-nearest neighbor finding using MaxNearestDist. IEEE Trans Pattern Anal Mach Intell 30(2):243–252
13. Suthaharan S, Suthaharan S (2016) Support vector machine. In: Machine learning models and algorithms for big data classification: thinking with examples for effective learning, pp 207–235
14. De Ville B (2013) Decision trees. Wiley Interdiscip Rev: Comput Stat 5(6):448–455
15. Schulz E, Speekenbrink M, Krause A (2018) A tutorial on Gaussian process regression: modelling, exploring, and exploiting functions. J Math Psychol 85:1–16
16. Artrith N, Kolpak AM (2015) Grand canonical molecular dynamics simulations of Cu–Au nanoalloys in thermal equilibrium using reactive ANN potentials. Comput Mater Sci 110:20–28
17. Li Z, Meidani K, Yadav P, Barati Farimani A (2022) Graph neural networks accelerated molecular dynamics. J Chem Phys 156(14)
18. Peivaste I, Ramezani S, Alahyarizadeh G, Ghaderi R, Makradi A, Belouettar S (2024) Rapid and accurate predictions of perfect and defective material properties in atomistic simulation using the power of 3D CNN-based trained artificial neural networks. Sci Rep 14(1):36
19. Dietterich TG (2000) Ensemble methods in machine learning. In: International workshop on multiple classifier systems. Springer Berlin Heidelberg, Berlin, Heidelberg, pp 1–15
20. Kadiyala A, Kumar A (2018) Applications of python to evaluate the performance of bagging methods. Environ Prog Sustain Energy 37(5):1555–1559
21. Duffy N, Helmbold D (2002) Boosting methods for regression. Mach Learn 47:153–200
22. Chen T, Guestrin C (2016) Xgboost: a scalable tree boosting system. In: Proceedings of the 22nd ACM sigkdd international conference on knowledge discovery and data mining, pp 785–794
23. Saha S, Kumar Gupta K, Ranjan Maity S, Dey S (2023) Probabilistic investigation of geometric responses in Wire EDM machined complex-shaped profile: a machine learning based approach. Proc Inst Mech Eng Part B: J Eng Manuf 237(12):1798–1809
24. Gaussian Process Regression (GPR) GeeksforGeeks. Last updated 24 April2025
25. Sankar A, Gupta KK, Bhalla V, Pandey DS (2025) Multi-criteria optimization of nanofluid-based solar collector for enhanced performance: an explainable machine learning-driven approach. Energy 320:135212

Chapter 4
Prospects of Machine Learning-Driven Atomistic Simulations: A Review

4.1 Overview

Over the centuries, the materials exploration has greatly relied upon intuition, large-scale iterative experimentation, and, sometimes, serendipitous innovations. The recent advent of artificial intelligence (AI) and recent advancements in computational technology create a strong rationale for coming up with a structured and robust framework to material design and discovery. Notably, AI has emerged as a potential tool, accelerating the identification of anomalies, predicting stable compounds, and forecasting material properties. This shift towards computational screening has proved highly beneficial by significantly cutting costs, saving time, and conserving resources, particularly in the early stages of material selection. Motivated by these advancements, this chapter delves into key areas of atomistic simulations where the integration of machine learning (ML) has had a transformative effect on materials research.

Machine learning (ML) is rapidly becoming an indispensable part of materials science, spanning diverse areas such as data-driven computational simulations, modelling physical phenomena, and even drug development [1]. Moving beyond traditional methods, the adoption of informatics-driven strategies has exceptionally accelerated the preliminary screening of material systems. With recent improvements in computational power, atomistic machine learning has become a powerful approach to exploring the fundamental physics of materials at the atomic level—especially when they are subjected to extreme conditions. While atomistic simulations provide precise and effective means to characterize materials at the nanoscale, combining ML with these simulations offers a more holistic understanding of material behaviour—an insight that is often constrained by the high computational costs associated with large-scale molecular dynamics (MD) simulations. The successful integration of ML models with MD simulations is paving the way for the development of automated frameworks in material characterization [2]. These frameworks are now capable of unveiling the complex interactions between processing, structure, and properties, an

K. K. Gupta et al., *Machine Learning in Nanoscale Materials Design*, Materials Horizons: From Nature to Nanomaterials, https://doi.org/10.1007/978-981-95-2660-4_4

ongoing challenge within the materials science community. The following sections will explore the promising areas where atomistic simulations can benefit from the integration of machine learning.

4.2 Challenges in High-Throughput MD Simulations

The large-scale experiments in the non-scale regime require an expensive and sophisticated lab setting, abundant resources, and investment of time and human resources. Such impractical requirements most often hinder the seamless discovery of the materials system. With the growing computational capabilities, the research community is rapidly turning towards computational simulation-driven preliminary screening—particularly in nanoscale investigations. This transition has led to the widespread use of molecular dynamics (MD) simulations over the past few decades. Even though MD simulations are efficient, performing such simulations for large-scale sample space is computationally expensive. Moreover, performing MD simulations for larger systems (with number of atoms$\approx 10^6$) typically requires using supercomputing resources and long execution times, which can again hinder the pace of materials discovery and limit the depth of insights into their mechanical and multi-physical behaviours.

The recent progress in machine learning has introduced new avenues for exploring nanoscale materials. Artificial intelligence offers the potential to push the boundaries of material behaviour predictions more effectively. The integration of machine learning techniques with computational materials science has notably advanced the analysis and understanding of nanoscale material characteristics [3, 4]. Some research groups have begun using machine learning-supported statistical and probabilistic models as partial replacements for traditional molecular simulations. For example, Mukhopadhyay et al. [4] investigated how strain rate and temperature affect the yield strength of hexagonal close-packed (HCP) nanowires and proposed a machine learning-assisted surrogate modelling approach for performing detailed uncertainty analysis—one of the early efforts in this direction. The investigation reported by Garg et al. [5] explored changes in the thermal conductivity of graphene as a function of temperature and Stone–Wales defects, developing soft computing-based prediction models. Vijayraghavan et al. [6] used multi-gene genetic programming to define how compressive strength in BNNTs varies with aspect ratio, temperature, and vacancy defects. In a similar vein, Garg and Tai [7] employed a modified multi-gene genetic programming method incorporating stepwise regression to predict surface finish in turning processes. Garg et al. [8] developed an ML model to estimate the nano-machining performance of BNN panels by examining process parameters such as temperature, feed rate, and drill bit speed.

Yang et al. [9] applied several supervised learning models, including artificial neural networks, to estimate interfacial thermal resistance between graphene and hBN sheets. Wang et al. [10] implemented a supervised random forest model to predict the mechanical behaviour of tungsten disulphide, accounting for chirality, system size, temperature, strain rate, and vacancy defects. Zhang et al. [11] introduced a machine learning framework integrated with molecular simulations, where training and testing were conducted on a significantly large dataset ($N = 1440$).

This brief overview suggests that the scientific community is actively examining the influence of both internal and external factors on the mechanical properties of nanoscale materials using MD simulations. These efforts often incorporate efficient probabilistic approaches to obtain detailed insights. High-throughput atomistic simulations are especially valuable, as they reduce the time and resources required for initial material screening in targeted applications [12]. The reviewed literature indicates that the high computational demands of full-scale MD simulations can be effectively addressed by combining efficient sampling techniques—such as Monte Carlo sampling, which captures broad parametric variability—with machine learning methods [13–18] (refer to Fig. 4.1). Subsequent chapters of this book explore similar strategies in greater depth, offering a detailed look at how machine learning can be integrated into conventional MD workflows for large-scale predictions, virtual screening, and other downstream applications.

4.3 Machine Learning-Based Interatomic Potential (MLIPs)

Molecular dynamics (MD) simulations have become an essential tool for analysing nanoscale materials. However, the accuracy of the results from these simulations relies heavily on the interatomic potential (IP) used, which is mathematically calibrated based on data from first-principles methods or experiments [19, 20].

In general, the interatomic potentials are mathematical representations of the structure (interatomic distance) and stability (potential energy) relationship, which is fitted by performing large-scale in-lab experiments or quantum-scale simulations (refer to Fig. 4.2). The utilization of ML in the development of interatomic potential has seen a great surge due to the resource and time intensiveness involved in physical experiments and simulations [21]. For example, Thong et al. [22] utilized the dataset from quantum mechanical simulations to train the deep neural network for developing the interatomic potential to model ferroelectric perovskites ($KNbO_3$). Similarly, Tang et al. [23] utilized a neural network to learn the interactions between Al–Ce in metallic liquid alloys. They utilized ANN-based interatomic potential for performing the MD simulations for Al–Ce compounds. Ladygin et al. [24] used the density functional theory calculation to train the ML interatomic potential to model the composition of four different metallic elements (Al, Mo, Ti, and U). To bridge the gap between quantum-scale and nanoscale simulations, Wang et al. [25] developed MLIP by using ab initio MD (AIMD) simulations. Shayestehpour and Zahn [26] utilized the DFT calculations and machine learning model to model the interatomic potential for deep eutectic solvents. Such utilization of ML models in development of the machine learning-based interatomic potentials (MLIPs) opens up new avenues for computationally characterizing the materials systems for which

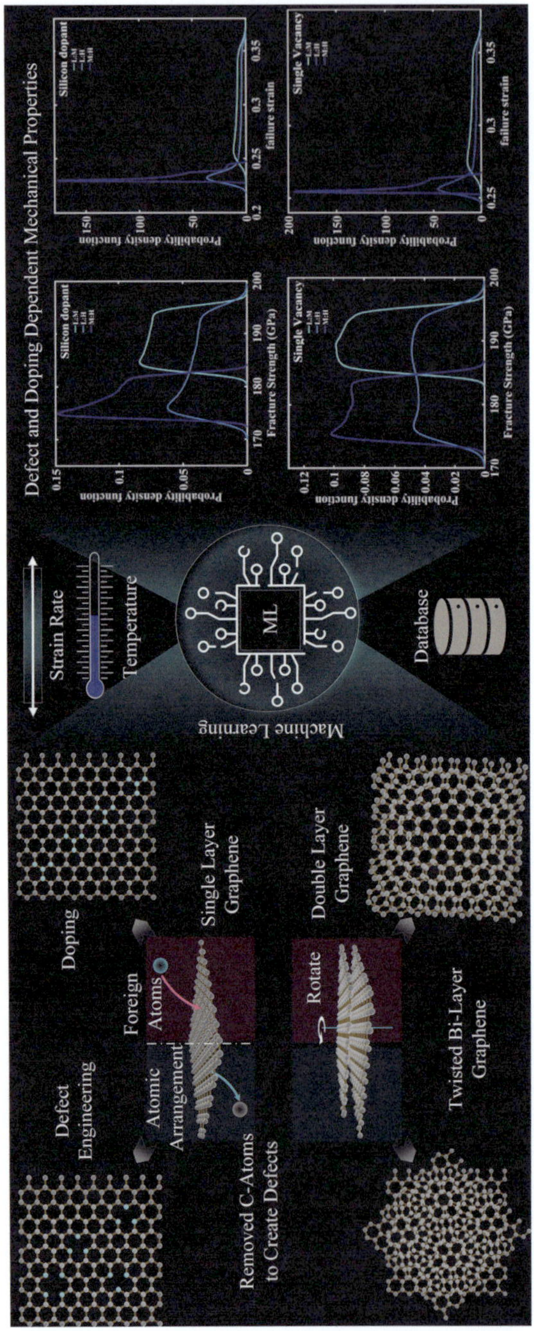

Fig. 4.1 Machine learning-driven molecular dynamics simulation [13–18]

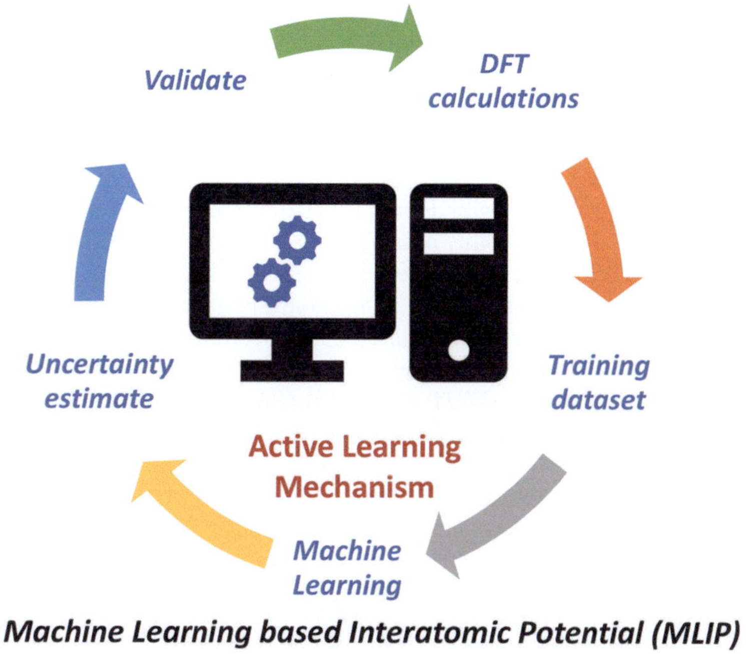

Fig. 4.2 Machine learning-based interatomic potential [20]

the interatomic potentials are not currently available. Datasets are the foundation of machine learning, serving as the essential input from which algorithms learn and generate insights. Diverse and comprehensive datasets contribute to the development of robust models that can generalize effectively across a range of scenarios, thereby minimizing bias and enhancing predictive accuracy. In the field of materials science, gathering rich datasets through computationally intensive methods, such as density functional theory (DFT) calculations, is crucial for capturing the full range of material properties and behaviours [27].

Such extensive data collection not only aids in discovering new materials with targeted properties but also deepens our understanding of complex material interactions and enables reliable predictions of material performance under varying conditions. The available open-source materials databases, such as ICSD [28], Materials Project [29], Aflow [30], COD [31], and C2DB [32], help in getting the dataset for training the machine learning interatomic potentials (MLIPs).

4.4 Nanoscale Design of Materials

Materials-by-design is a new-generation approach which aims at accelerating the design and discovery of function-specific material by algorithmically exploring the large sample space during computational screening. This approach leverages machine learning methods and theoretical calculations (numerical simulations, molecular dynamics simulations, density functional theory, etc.) to reduce the timeline for materials discovery from decades to months (refer to Fig. 4.3). The complex relationship among a material's processing, structure, and properties (PSP) drives its functional performance. Strategically modifying a material's attributes (such as structure and composition) at the nano- and microscale can enhance its performance [33–38]. Exploring materials with desired properties through manual experimentation is hindered by the infeasibility of large-scale experimentation needed to understand complex process–structure–property relationships. Materials-by-design offers the solution in terms of implementing AI-integrated computational simulations for exploring and controlling the relationship among structure, processing, composition, and properties. The successful implementation of these concepts has shown promise in revealing the deep insights of PSP linkages; for instance, Liu et al. [39] deployed machine learning for establishing the relationship among processing, structure, and mechanical performance of additively manufactured metal bars. Thoppil et al. [40] deployed a Bayesian framework to establish PSP linkages of complex concentrated alloys by exploring the thermo-mechanical performance dataset. Tran et al. [41] developed an ML-driven predictive framework for rapid characterization of polymers in terms of thermal, mechanical, and electrical performance. The integration of AI/ML in materials research has shown great promise in terms of accelerated search of materials with desired performance.

4.5 Summary

This chapter explores key areas within atomistic simulations where the integration of machine learning models can provide more efficient and timely solutions. These models are particularly useful in overcoming challenges such as the high computational costs and time demands of conventional simulations, the scarcity of detailed morphological and stability data (through interatomic potentials), and the complexities associated with designing nanoscale materials. Subsequent chapters expand on these ideas, providing practical insights and strategies to address some of the significant physical challenges discussed here.

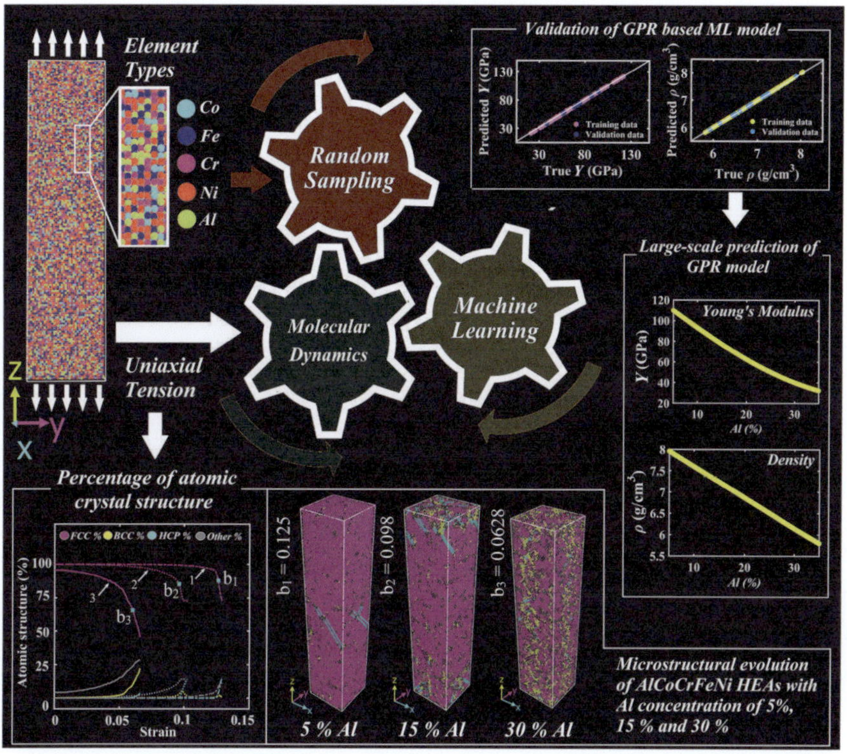

Fig. 4.3 Nanoscale design of AlCoCrFeNi HEA by deploying the machine learning-driven optimization framework [42]

References

1. Margraf JT (2023) Science-driven atomistic machine learning. Angew Chem Int Ed 62(26):e202219170
2. Axelrod S, Schwalbe-Koda D, Mohapatra S, Damewood J, Greenman KP, Gómez-Bombarelli R (2022) Learning matter: materials design with machine learning and atomistic simulations. Acc Mater Res 3(3):343–357
3. Yu CH, Qin Z, Buehler MJ (2019) Artificial intelligence design algorithm for nanocomposites optimized for shear crack resistance. Nano Futures 3(3):035001
4. Mukhopadhyay T, Mahata A, Dey S, Adhikari S (2016) Probabilistic analysis and design of HCP nanowires: an efficient surrogate based molecular dynamics simulation approach. J Mater Sci Technol 32(12):1345–1351
5. Garg A, Vijayaraghavan V, Wong CH, Tai K, Gao L (2014) An embedded simulation approach for modelling the thermal conductivity of 2D nanoscale material. Simul Model Pract Theory 44:1–13
6. Vijayaraghavan V, Garg A, Wong CH, Tai K, Singru PM (2015) An integrated computational approach for determining the elastic properties of boron nitride nanotubes. Int J Mech Mater Des 11(1):1–14
7. Garg A, Tai K (2014) Stepwise approach for the evolution of generalized genetic programming model in prediction of surface finish of the turning process. Adv Eng Softw 78:16–27

8. Garg A, Vijayaraghavan V, Lam JSL, Singru PM, Gao L (2015) A molecular simulation based computational intelligence study of a nano-machining process with implications on its environmental performance. Swarm Evol Comput 21:54–63
9. Yang H, Zhang Z, Zhang J, Zeng XC (2018) Machine learning and artificial neural network prediction of interfacial thermal resistance between graphene and hexagonal boron nitride. Nanoscale 10(40):19092–19099
10. Wang X, Han D, Hong Y, Sun H, Zhang J, Zhang J (2019) Machine learning enabled prediction of mechanical properties of tungsten disulfide monolayer. ACS Omega 4(6):10121–10128
11. Zhang Z, Hong Y, Hou B, Zhang Z, Negahban M, Zhang J (2019) Accelerated discoveries of mechanical properties of graphene using machine learning and high-throughput computation. Carbon 148:115–123
12. Erucar I, Keskin S (2018) High-throughput molecular simulations of metal organic frameworks for CO_2 separation: opportunities and challenges. Front Mater 5:4
13. Gupta KK, Mukhopadhyay T, Roy A, Roy L, Dey S (2021) Sparse machine learning assisted deep computational insights on the mechanical properties of graphene with intrinsic defects and doping. J Phys Chem Solids 155:110111
14. Gupta KK, Mukhopadhyay T, Roy L, Dey S (2022) High-velocity ballistics of twisted bilayer graphene under stochastic disorder. Adv Nano Res 12(5):529–547
15. Gupta KK, Roy L, Dey S (2022) Hybrid machine-learning-assisted stochastic nano-indentation behaviour of twisted bilayer graphene. J Phys Chem Solids 167:110711
16. Gupta KK, Roy A, Mukhopadhyay T, Roy L, Dey S (2022) Probing the stochastic fracture behavior of twisted bilayer graphene: efficient ANN based molecular dynamics simulations for complete probabilistic characterization. Mater Today Commun 32:103932
17. Gupta KK, Roy L, Dey S (2022) Ballistic performance of bi-layer graphene: artificial neural network based molecular dynamics simulations. In: Machine learning applied to composite materials. Springer Nature Singapore, Singapore, pp 135–148
18. Gupta KK, Roy L, Dey S (2022) Machine learning-based molecular dynamics simulations of monolayered graphene. In: Recent advances in computational and experimental mechanics, Vol II: select proceedings of ICRACEM 2020. Springer Nature Singapore, Singapore, pp 251–263
19. Gupta KK, Mukhopadhyay T, Roy L, Dey S (2022) Hybrid machine-learning-assisted quantification of the compound internal and external uncertainties of graphene: towards inclusive analysis and design. Mater Adv 3(2):1160–1181
20. Kulichenko M, Nebgen B, Lubbers N, Smith JS, Barros K, Allen AE, Habib A, Shinkle E, Fedik N, Li YW, Messerly RA (2024) Data generation for machine learning interatomic potentials and beyond. Chem Rev 124(24):13681–13714
21. Chan H, Narayanan B, Cherukara MJ, Sen FG, Sasikumar K, Gray SK, Chan MK, Sankaranarayanan SK (2019) Machine learning classical interatomic potentials for molecular dynamics from first-principles training data. J Phys Chem C 123(12):6941–6957
22. Thong HC, Wang X, Han J, Zhang L, Li B, Wang K, Xu B (2023) Machine learning interatomic potential for molecular dynamics simulation of the ferroelectric KNb O3 perovskite. Phys Rev B 107(1):014101
23. Tang L, Ho KM, Wang CZ (2021) Molecular dynamics simulation of metallic Al–Ce liquids using a neural network machine learning interatomic potential. J Chem Phys 155(19)
24. Ladygin VV, Korotaev PY, Yanilkin AV, Shapeev AV (2020) Lattice dynamics simulation using machine learning interatomic potentials. Comput Mater Sci 172:109333
25. Wang P, Shao Y, Wang H, Yang W (2018) Accurate interatomic force field for molecular dynamics simulation by hybridizing classical and machine learning potentials. Extreme Mech Lett 24:1–5
26. Shayestehpour O, Zahn S (2023) Efficient molecular dynamics simulations of deep eutectic solvents with first-principles accuracy using machine learning interatomic potentials. J Chem Theory Comput 19(23):8732–8742
27. Wang G, Wang C, Zhang X, Li Z, Zhou J, Sun Z (2024) Machine learning interatomic potential: bridge the gap between small-scale models and realistic device-scale simulations. Iscience

28. Belsky A, Hellenbrandt M, Karen VL, Luksch P (2002) New developments in the inorganic crystal structure database (ICSD): accessibility in support of materials research and design. Struct Sci 58(3):364–369
29. Jain A, Ong SP, Hautier G, Chen W, Richards WD, Dacek S, Cholia S, Gunter D, Skinner D, Ceder G, Persson KA (2013) Commentary: the materials project: a materials genome approach to accelerating materials innovation. APL Mater 1(1)
30. Curtarolo S, Setyawan W, Hart GL, Jahnatek M, Chepulskii RV, Taylor RH, Wang S, Xue J, Yang K, Levy O, Mehl MJ (2012) AFLOW: an automatic framework for high-throughput materials discovery. Comput Mater Sci 58:218–226
31. Quirós M, Gražulis S, Girdzijauskaitė S, Merkys A, Vaitkus A (2018) Using SMILES strings for the description of chemical connectivity in the crystallography open database. J Cheminform 10(1):23
32. Haastrup S, Strange M, Pandey M, Deilmann T, Schmidt PS, Hinsche NF, Gjerding MN, Torelli D, Larsen PM, Riis-Jensen AC, Gath J (2018) The computational 2D materials database: high-throughput modeling and discovery of atomically thin crystals. 2D Mater 5(4):042002
33. Barman S, Kumar Gupta K, Dey S (2024) Exploring nano-scale scratching induced tribological behavior of graphene engineered AlCoCrFeNi high-entropy alloy. J Appl Mech 91(10)
34. Barman S, Gupta KK, Dey S (2024) Probing atomistic deformation behavior of graphene-coated Al0. 3CoCrFeNi high-entropy alloy under nanoindentation. J Micromech Molec Phys:1–14
35. Barman S, Gupta KK, Dey S (2024) Enhancing mechanical performance of Al0. 3CoCr-FeNi HEA films through graphene coating: insights from nanoindentation and dislocation mechanism analysis. Modell Simul Mater Sci Eng 32(3):035012
36. Roy A, Gupta KK, Dey S, Mukhopadhyay T (2024) Random topological defects in double-walled carbon nanotubes: on characterization and programmable defect-engineering of spatio-mechanical properties. Adv Nano Res 16(1):91
37. Gupta KK, Mukhopadhyay T, Dey S (2023) Probing the molecular-level energy absorption mechanism and strategic sequencing of graphene/Al composite laminates under high-velocity ballistic impact of nano-projectiles. Appl Surf Sci 629:156502
38. Gupta KK, Mukhopadhyay T, Roy A, Dey S (2020) Probing the compound effect of spatially varying intrinsic defects and doping on mechanical properties of hybrid graphene monolayers. J Mater Sci Technol 50:44–58
39. Liu Q, Chen W, Yakubov V, Kruzic JJ, Wang CH, Li X (2024) Interpretable machine learning approach for exploring process-structure-property relationships in metal additive manufacturing. Addit Manuf 85
40. Thoppil GS, Nie JF, Alankar A (2023) Bayesian approach for inferrable machine learning models of process–structure–property linkages in complex concentrated alloys. J Alloy Compd 967:171595
41. Tran H, Kim C, Gurnani R, Hvidsten O, DeSimpliciis J, Ramprasad R, Gadelrab K, Tuffile C, Molinari N, Kitchaev D, Kornbluth M (2024) Polymer composites informatics for flammability, thermal, mechanical and electrical property predictions. arXiv preprint arXiv:2412.08407
42. Gupta KK, Barman S, Dey S, Mukhopadhyay T (2024) Explainable machine learning assisted molecular-level insights for enhanced specific stiffness exploiting the large compositional space of AlCoCrFeNi high entropy alloys. Mach Learn: Sci Technol 5(2):025082

Chapter 5
Fracture Response of Graphene: A Data-Driven Characterization

5.1 Overview

The discovery and synthesis of graphene have sparked considerable interest across the scientific community, particularly due to its potential in a wide range of applications—from flexible and wearable electronics to energy storage systems, battery components, and composite materials. Over the last decade, this enthusiasm has driven the development of numerous nanoelectromechanical systems (NEMS) tailored for specific functions. However, achieving defect-free graphene remains a significant challenge. The production process, which often involves high temperatures and reactive chemical environments, tends to introduce structural imperfections. These can appear at the atomic level—such as vacancies, bond rotations, and nanopores—or as multi-atom inclusions like silicon, boron, or nitrogen dopants. Researchers have thoroughly investigated how these irregularities impact the mechanical behaviour of graphene, and the findings consistently show that such defects play a crucial role in altering its strength and stiffness. As a result, reliably predicting the mechanical performance of doped or defect-laden graphene is essential, especially for applications where structural integrity and deformation response are critical. In this chapter, a machine learning-driven framework is employed to explore how fracture strength in graphene varies with temperature and strain rate. The study focuses on several configurations, including pristine graphene, graphene with a single vacancy, graphene containing a silicon dopant, and graphene featuring both a vacancy and silicon doping.

The exceptional functional attributes offered by graphene monolayers and their heterostructures have led to their successful utilization in the development of wide varieties of nanoelectromechanical systems (NEMS) [1] for the applications in sensing mechanism [2], gas/liquid separation [3], energy storage [4], composite materials [5], etc. Following the successful synthesis of monolayer graphene [6], it has rapidly gained the interest for both theoretical and experimental research, owing to

K. K. Gupta et al., *Machine Learning in Nanoscale Materials Design*, Materials Horizons: From Nature to Nanomaterials, https://doi.org/10.1007/978-981-95-2660-4_5

the potential solution it can offer in various domains. Since its discovery, significant efforts have been made to investigate the mechanical and fracture behaviour of monolayer graphene sheets. It is now common knowledge that graphene is among the lightest materials known (0.77 mg/m^2) while also exhibiting exceptional stiffness. The fracture strength and Young's modulus of graphene are measured as 130 GPa and 1 TPa, respectively [7–9]. Such remarkable mechanical characteristics have opened up novel avenues for incorporating graphene into composite materials, where it has the potential of serving as a reinforcing agent to enhance structural performance [10]. In general, there are multiple techniques available to produce graphene, with considerations for cost-efficiency, quality, and production volume. Common top-down synthesis methods include micromechanical cleavage, arc discharge, unrolling of carbon nanotubes, chemical vapour deposition (CVD), and epitaxial growth on silicon carbide (SiC). Among these, CVD stands out as a viable option for commercial-scale production. However, none of these methods guarantees the creation of defect-free graphene. The involvement of chemical processes and high temperatures during production often introduces various structural imperfections and dopants into the material. Extensive research has focused on understanding the intrinsic properties of pristine graphene, but it is clear that production-induced topological defects—largely unavoidable—have a significant influence on its mechanical behaviour. The practical use of graphene hinges on the ability to mitigate the negative impact of these structural anomalies. These effects are further compounded by variations in environmental factors such as temperature and strain rate [11].

Thus, it is critical to analysing the combined effects of defect types, dopant levels, temperature, and strain rate. Experimental work at the nanoscale, however, faces challenges in terms of accuracy, cost, and time. With increasing computational capabilities, researchers have increasingly turned to simulations as a partial alternative to laboratory experiments, particularly for nanoscale systems. This shift has led to a notable rise in the use of molecular dynamics (MD) simulations over recent decades. Still, one of the main drawbacks of MD simulations lies in their high computational demand and time requirements, which pose challenges to conducting comprehensive analyses of nanoscale systems. Large-scale MD simulations typically require access to supercomputing resources and can be prohibitively time-consuming, limiting the pace of material discovery and the depth of computational understanding of their mechanical and multi-physical properties. These limitations have fuelled a growing interest in integrating artificial intelligence to complement existing simulation practices and extend the reach of molecular-level insights.

In this chapter, a machine learning (ML)-based predictive framework is developed to perform large-scale prediction of the fracture strength of graphene by capturing the underlying physics of the system by running an optimal minimum number of molecular simulations based on a D-optimal design algorithm. Five different (configurational and operational) input features such as concentration of single vacancy (SV) defects, concentration of silicon atom dopants (Si), edge effect (Edge), temperature (T), and strain rate (SR), which influence the critical mechanical behaviour of monolayer graphene, would be investigated for fracture strength.

5.2 MD Simulation-Driven Dataset

For the generation of initial sample space, a design of experiments (DOE) strategy is utilized [12], which resulted in 288 samples. The full-factorial design is implemented by using the practically relevant (from the published literature) input features in terms of discrete levels (as shown in Table 5.1).

The full-factorial design of samples is performed by using the *itertools* package of the Python environment by using the code snippet illustrated in Fig. 5.1. The full-factorial design of sample space provided all the possible combinations of the levels of the features used in the investigation, and resulted in 288 input feature settings for performing the MD simulations of uniaxial tensile deformation of graphene configurations by varying temperature and strain rate.

By enforcing varied levels of single vacancy concentrations, and silicon dopant concentrations, the presented investigation mainly explores the fracture response of four different configurations of graphene (refer to Fig. 5.2) subjected with varying temperature and strain rate conditions.

The challenge with full-factorial design-based sample space is that the input features are distributed in terms of discrete levels. To make it continuous, 16 random samples (consisting of the input feature value in between the parametric range of variation) are added to the sample space obtained from the full-factorial design. In total 304 (288 + 16) simulations were performed by using the LAMMPS, and as an outcome of the simulations fracture strength is recorded through a stress–strain profile obtained from an individual simulation.

Although molecular simulations provide high-fidelity information for all 304 samples, we currently have a majority of the information for the discrete levels only, which were used to initiate the study. The behaviour of graphene—whether pristine, defective, doped, or both defective and doped—remains unknown within the temperature ranges of 100–300 K, 300–600 K, and 600–900 K. Similarly, there are limitations in understanding the strain rate-dependent elastic response of graphene and its behaviour under different loading directions. This challenge can be addressed by developing an efficient machine learning (ML) model capable of accurately

Table 5.1 Different levels of input features used in the analysis based on D-optimal design

Features	Level-1	Level-2	Level-3	Level-4
Temperature (T) (K)	100	300	600	900
Strain rate (SR) (/ps)	0.0005	0.001	0.0025	0.005
Single vacancy concentration (SV) (%)	0	0.5	1	–
Silicon dopant concentration (Si) (%)	0	0.5	1	–
Edge effect (Edge)	Armchair	Zigzag	–	–

Four different levels are considered for the external parameters, while three levels are considered for the internal input features with numerical values and two levels are there for the categorical variable edge effect [11]

```
import itertools
import pandas as pd

# Define factor levels
factors = {
    "Temperature (K)": [100, 300, 600, 900],
    "Strain rate (/ps)": [0.0005, 0.001, 0.0025, 0.005],
    "Single Vacancy concentration (%)": [0, 0.5, 1],
    "Silicon dopant concentration (%)": [0, 0.5, 1],
    "Edge effect": ["Armchair", "Zigzag"]
}

# Generate full factorial design
factorial_design = list(itertools.product(*factors.values()))

# Convert to DataFrame
df = pd.DataFrame(factorial_design, columns=factors.keys())

# Display the first few rows
print(df.head())

# Save to CSV
df.to_csv("full_factorial_design.csv", index=False)
```

Fig. 5.1 Snippet of the Python code utilized for full-factorial design of DOE

predicting the fracture strength of various graphene configurations for the continuous variations in temperature and strain rate.

5.3 Machine Learning Pipeline

The machine learning (ML) pipeline starts with the preprocessing of the dataset, which involves one-hot encoding of the categorical variables (such as Edge) and normalization of the continuous variables (such as temperature (T), strain rate (SR), SV(%), and Si (%)). One-hot encoding of the categorical variables ensures that the features are not treated as continuous variables during model formation, whereas normalization (z-score) of continuous variables ensures that there is no scale difference. For instance, in the considered dataset temperature varies from 100 to 900 K, while strain rate varies from 0.0005 to 0.005/ps, which shows that there is a significant scale difference among the variables T and SR. Z-score normalization is performed by using the following formulation:

$$X_i\_\text{normalized} = \frac{(X_i - \mu(X_i))}{\sigma(X_i)} \tag{5.1}$$

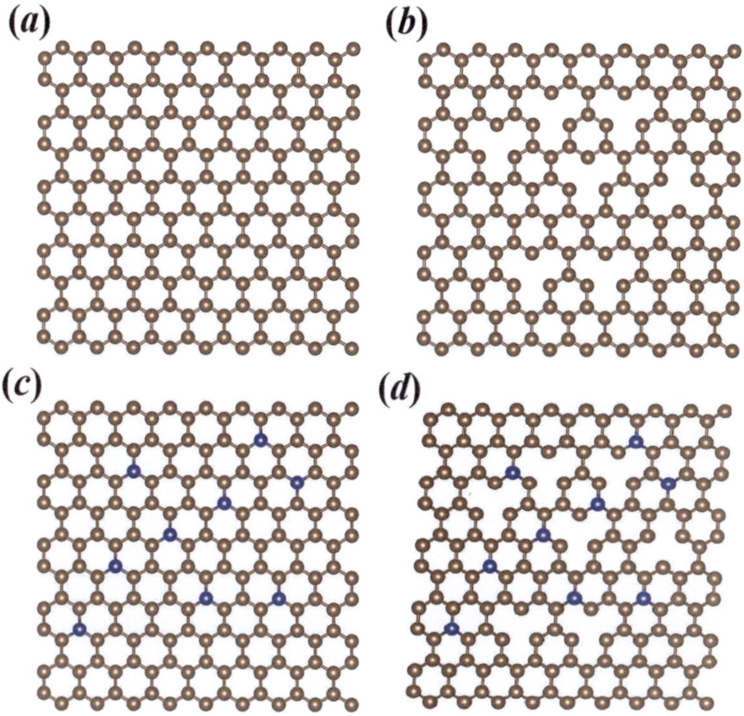

Fig. 5.2 Different configurations of graphene monolayers considered in the investigation. **a** Pristine graphene, **b** graphene subjected with single vacancy defects, **c** graphene subjected with silicon (Si) doping, and **d** graphene subjected with single vacancy defects and silicon (Si) doping, simultaneously

The ML pipeline in Python is executed by using *scikit-learn* library; in the following part of this section, the construction and validation of the ML pipeline in the Python environment are chronologically laid out.

5.3.1 Data Preprocessing

After importing the essential Python dependencies such as numpy, pandas, and matplotlib.pyplot, the data preprocessing is performed, which consists of

- Identifying input features and response from the dataset, random train–test–validation split (10% test, 10% validation) (Python snippet shown in Fig. 5.3)
- One-hot encoding for categorical input features such as 'Edge', and z-score normalization for continuous numerical input features, such as 'Temperature', 'sR', 'sV', and 'si' (refer to Python code snippet shown in Fig. 5.4).

```
import pandas as pd
import numpy as np
import seaborn as sns
import matplotlib.pyplot as plt
from sklearn.model_selection import train_test_split
# Load your dataset from Excel
file_path = '/content/Dataset.xlsx'
df = pd.read_excel(file_path)

# Indentifying Input features (X), and Response(y)
X = df[['Temperature', 'SR', 'SV', 'Si', 'Edge']]
y = df['Fstrength']

# Train-test-validation split
from sklearn.model_selection import train_test_split
X_temp, X_test, y_temp, y_test = train_test_split(X,
                                                  y,
                                                  test_size=0.1,
                                                  random_state=42)
X_train, X_val, y_train, y_val = train_test_split(X_temp,
                                                  y_temp,
                                                  test_size=0.1,
                                                  random_state=42)
```

Fig. 5.3 Python snippet for importing dataset, extracting input features and target, and random train–test split

```
# Identify categorical features, Normalization of Continuous features
from sklearn.preprocessing import OneHotEncoder
from sklearn.compose import ColumnTransformer
from sklearn.pipeline import Pipeline
from sklearn.preprocessing import StandardScaler

categorical_features = ['Edge']

# Preprocessing categorical features with one-hot encoding
# Preprocessing continuous features with StandarScaler (z-score)
preprocessor = ColumnTransformer([
    ('onehot', OneHotEncoder(handle_unknown='ignore'), categorical_features),
    ('scale', StandardScaler(), ['Temperature', 'SR', 'SV', 'Si'])
])

# Transform input features
X_train_transformed = preprocessor.fit_transform(X_train)
X_test_transformed = preprocessor.transform(X_test)
X_val_transformed = preprocessor.transform(X_val)
```

Fig. 5.4 Python snippet for data preprocessing: one-hot encoding for categorical variables and z-score normalization for continuous variables

5.3.2 Selection of ML Models

Selecting the right machine learning (ML) model can be challenging, as there is no one-size-fits-all approach. A model's effectiveness and predictive accuracy depend on several factors, such as the nature of the data, the size and quality of the dataset, and feature engineering. When it's uncertain which ML model will be most effective for a given dataset, using Python's *LazyPredict* package is a great option. It enables quick testing and comparison of a wide range of models (*e.g.* linear models, decision trees, support vector machines, and random forests) and provides a performance summary for each one. This tool is particularly valuable during the early stages of a project, as it provides a fast overview of which models perform best. It saves time by eliminating the need to manually train and evaluate each model individually. In the following stage, the *LazyPredict* package is used to provide the models' performance summary for all the responses. The Python snippet for performing *LazyPredict* evaluation is illustrated in Fig. 5.5.

The comparison of the ML models offered by *LazyPredict* revealed that the tree-based ensemble ML models are relatively superior in making accurate predictions for fracture strength. The assessment of the model's performance is made by comparing the R-squared (R^2) values and root mean-squared error (RMSE); the observations offered by *LazyPredict* are illustrated in Fig. 5.6, which shows that gradient boosting regressor (GBR) and LGBM regressor are the best ML models for making the predictions of fracture strength. In the following investigation, these two models are selected, hyperparameter-tuned, and tested for out-of-the-fold samples.

```
# Perform lazypredict
!pip install lazypredict

import lazypredict
from lazypredict.Supervised import LazyRegressor

# Initialize LazyRegressor
lazy_reg = LazyRegressor()

# Fit and evaluate models for y
y_models, y_preds = lazy_reg.fit(X_train_transformed,
                                 X_test_transformed,
                                 y_train,
                                 y_test)
print("Results for Fstrength:\n", y_models)
#plot for R2 and RMSE heatmaps
fig, axes = plt.subplots(1, 2, figsize=(18, 6), dpi=300)
sns.heatmap(y_models[['R-Squared']],
            annot=True, cmap='Blues',
            fmt='.2f',
            linewidths=0.5,
            cbar_kws={'shrink': 0.8},
            ax=axes[0], vmin=-1, vmax=1)
axes[0].set_title("LazyPredict Model R2 Scores Heatmap", fontsize=16, fontweight='bold')
axes[0].tick_params(axis='x', rotation=45)
axes[0].tick_params(axis='y', labelsize=12)

sns.heatmap(y_models[['RMSE']],
            annot=True,
            cmap='Reds',
            fmt='.2f',
            linewidths=0.5,
            cbar_kws={'shrink': 0.8},
            ax=axes[1])
axes[1].set_title("LazyPredict Model RMSE Scores Heatmap", fontsize=16, fontweight='bold')
axes[1].tick_params(axis='x', rotation=45)
axes[1].tick_params(axis='y', labelsize=12)
```

Fig. 5.5 Python's *LazyPredict* for accelerated exploration of different ML model's performance

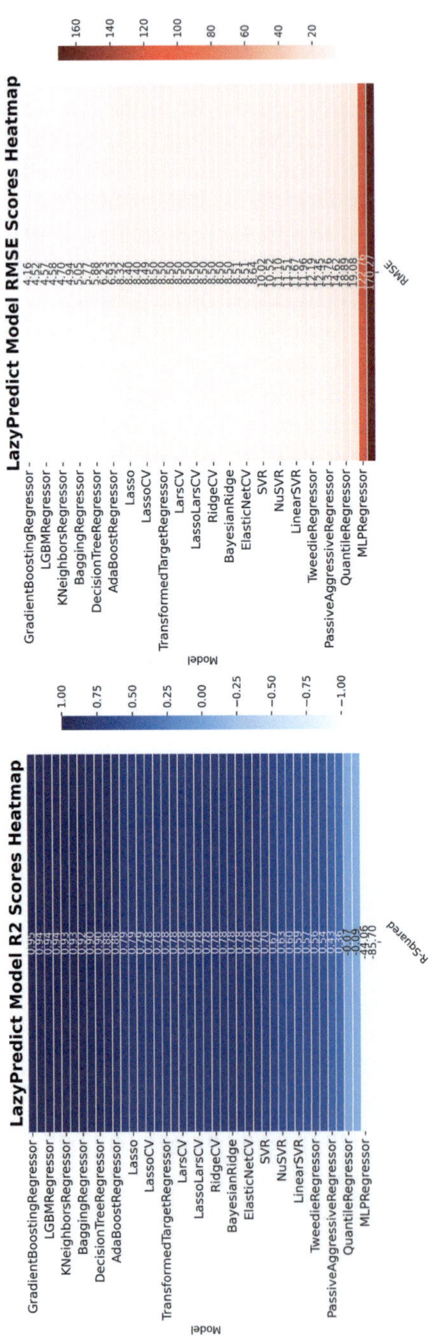

Fig. 5.6 Comparison of ML model performance offered by *LazyPredict*

5.3.3 Robustness of the Selected Models

The developed models so far utilized the training and test dataset to provide the preliminary screening of the applicability of ML models for the given dataset. In the following stage, the selected ML models (GBR and LGBM) will be rigorously trained and tested for their generalization capability. At first, k-fold ($k = 5$) cross-validation is applied with the hyperparameter tuning to develop the best possible ML models. Hyperparameters are essential components of machine learning models, and adjusting them can significantly impact the model's performance. This highlights the importance of optimally selecting hyperparameter values to develop the best possible model. The initialization of the selected models and the hyperparameters explored for the models are shown in the Python snippet illustrated in Fig. 5.7.

Next, a grid-search cross-validation mechanism (fivefold) is employed during model training to ensure the robustness of the developed ML models against dataset variability. The performance of the models is evaluated using the coefficient of determination (R^2) for each validation fold. This analysis is conducted for both Gradient Boosting Regressor (GBR) and LightGBM (LGBM), as demonstrated in the Python snippet provided in Fig. 5.8.

The cross-validation scores of both the ML models are illustrated in Fig. 5.9 in terms of a bar plot, which shows how the R^2 values vary across the 5 folds.

The best hyperparameters for both ML models, along with their corresponding R^2 values, are presented in the Python snippet shown in Fig. 5.10.

```python
from sklearn.ensemble import BaggingRegressor
from lightgbm import LGBMRegressor
from sklearn.model_selection import GridSearchCV, cross_val_score

# Define models
bagging_model = BaggingRegressor(random_state=42)  # Bagging Regressor
lgbm_model = LGBMRegressor(random_state=42)  # LGBM Regressor

# Hyperparameter tuning for Bagging Regressor
param_grid_bagging = {
    'n_estimators': [50, 100, 200],
    'max_samples': [0.5, 0.8, 1.0],  # fraction of samples to use for each base model
    'max_features': [0.5, 0.8, 1.0],  # fraction of features to use for each base model
}

# Hyperparameter tuning for LGBM Regressor
param_grid_lgbm = {
    'n_estimators': [50, 100, 200],
    'learning_rate': [0.01, 0.1, 0.2],
    'num_leaves': [31, 50, 100]
}
```

Fig. 5.7 Python snippet to initialize bagging regressor, and LGBM regressor by defining the hyperparameters to be tuned

```
# Grid search with cross-validation
bagging_grid = GridSearchCV(bagging_model, param_grid_bagging, cv=5, scoring='r2')
lgbm_grid = GridSearchCV(lgbm_model, param_grid_lgbm, cv=5, scoring='r2')

# Fit models
bagging_grid.fit(X_train_transformed, y_train)

lgbm_grid.fit(X_train_transformed, y_train)

# Evaluate models with 5-fold CV
bagging_cv_scores = cross_val_score(bagging_grid.best_estimator_,
                                    X_train_transformed,
                                    y_train, cv=5, scoring='r2')
lgbm_cv_scores = cross_val_score(lgbm_grid.best_estimator_,
                                 X_train_transformed,
                                 y_train, cv=5, scoring='r2')

print("Bagging Regressor CV Scores:", bagging_cv_scores)
print("LGBM Regressor CV Scores:", lgbm_cv_scores)
```

Fig. 5.8 Python snippet to perform grid-search *k*-fold cross-validation for hyperparameter tuning of the ML models

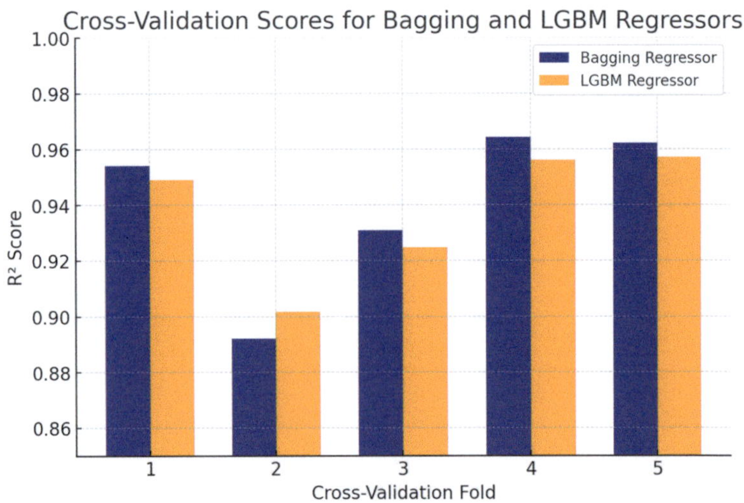

Fig. 5.9 Consistency of R^2 values across the fold of cross-validation for both the ML models

5.3.4 *Model Deployment for Prediction*

As evident from Figs. 5.9 and 5.10, the Bagging Regressor demonstrated superior predictive performance after hyperparameter tuning and *k*-fold cross-validation. Consequently, it is selected for making predictions on the unseen sample space (test and validation data).

The generalization capability of the Bagging Regressor—i.e. its ability to make accurate predictions on unseen data—is evaluated using R^2 values and error analysis. The latter includes a sample-wise distribution of relative errors (error/actual value).

```
print("Best Hyperparameters for Bagging Regressor:", bagging_grid.best_params_)
print("Best Score from Grid Search:", bagging_grid.best_score_)
print("Best Hyperparameters for LGBM Regressor:", lgbm_grid.best_params_)
print("Best Score from Grid Search:", lgbm_grid.best_score_)

Best Hyperparameters for Bagging Regressor: {'max_features': 1.0,
                                             'max_samples': 1.0,
                                             'n_estimators': 200}
Best Score from Grid Search: 0.9406393170513068
Best Hyperparameters for LGBM Regressor: {'learning_rate': 0.2,
                                          'n_estimators': 200,
                                          'num_leaves': 31}
Best Score from Grid Search: 0.9376572401880703
```

Fig. 5.10 Python snippet to print the best hyperparameters observed while hyperparameter tuning and recording the R^2 value for the observed best hyperparameters

```
# Predict on train, test, and validation data
y_pred_bagging_train = bagging_grid.best_estimator_.predict(X_train_transformed)
y_pred_bagging_test = bagging_grid.best_estimator_.predict(X_test_transformed)
y_pred_bagging_val = bagging_grid.best_estimator_.predict(X_val_transformed)

# Compute relative errors
relative_error_bagging_train = (y_pred_bagging_train - y_train) / y_train
relative_error_bagging_test = (y_pred_bagging_test - y_test) / y_test
relative_error_bagging_val = (y_pred_bagging_val - y_val) / y_val
```

Fig. 5.11 Python snippet to deploy the bagging regressor for making the predictions (train, test, and validation), and computing the relative error in the predictions

Two separate unseen datasets (test and validation) are used to assess the model's consistency across varying data distributions.

The Python snippet in Fig. 5.11 demonstrates the deployment of the Bagging Regressor for predicting fracture strength across the training, test, and validation datasets. These predictions are then compared with the actual values by evaluating the relative error, as illustrated in Fig. 5.11.

Next, the observations for training, test, and validation datasets are presented in terms of scatter plots, and error distribution plots by using the Python snippet illustrated in Fig. 5.12. The plotted observations are presented in Fig. 5.13.

The observations from Fig. 5.13 highlight the strong predictive capabilities of the trained Bagging Regressor, even when applied to completely unseen datasets (test and validation). The scatter plot in Fig. 5.13 shows that all sample points—across training, test, and validation datasets—follow a near-linear distribution with minimal spread. This indicates a low deviation between the predicted and actual fracture strength values, as recorded from MD simulations. Additionally, the error plots demonstrate that the relative errors for all datasets (training, test, and validation) are tightly distributed within the range of − 0.10 to 0.10. This minimal error distribution further underscores the exceptional generalization capability of the developed Bagging Regressor model.

```
# High-quality Bagging Regressor plots
fig, axes = plt.subplots(1, 2, figsize=(14, 6), dpi=300)
axes[0].scatter(y_train, y_pred_bagging_train, label='Training Data', alpha=0.7)
axes[0].scatter(y_test, y_pred_bagging_test, label='Test Data', alpha=0.7)
axes[0].scatter(y_val, y_pred_bagging_val, label='Validation Data', alpha=0.7)
axes[0].set_xlabel('Actual Fracture Strength', fontsize=12)
axes[0].set_ylabel('Predicted Fracture Strength', fontsize=12)
axes[0].set_title('Bagging Regressor: Actual vs Predicted Fracture Strength', fontsize=14)
axes[0].legend()
axes[0].grid(True)

sns.kdeplot(relative_error_bagging_train, label='Training', fill=True, ax=axes[1])
sns.kdeplot(relative_error_bagging_test, label='Test', fill=True, ax=axes[1])
sns.kdeplot(relative_error_bagging_val, label='Validation', fill=True, ax=axes[1])
axes[1].set_xlabel('Relative Error', fontsize=12)
axes[1].set_ylabel('Density', fontsize=12)
axes[1].set_title('Bagging Regressor: Distribution of Relative Errors', fontsize=14)
axes[1].legend()
axes[1].grid(True)
plt.tight_layout()
plt.show()
```

Fig. 5.12 Python snippet for illustrating the performance of bagging regressor in terms of scatter plots and error plots

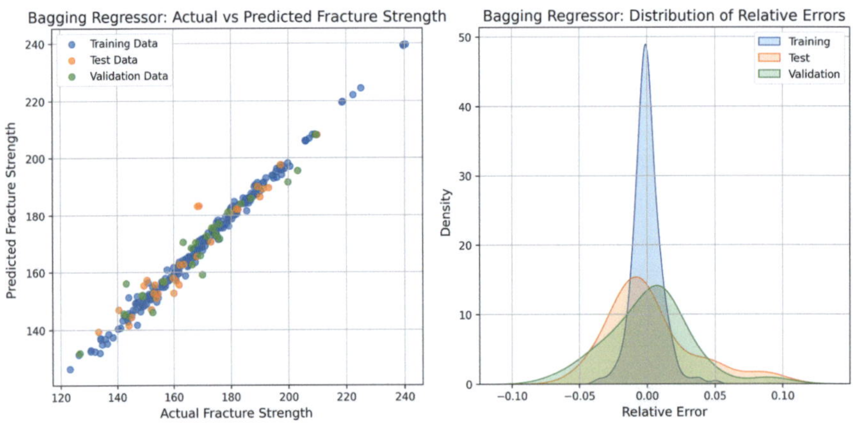

Fig. 5.13 Deployment of the Bagging Regressor for predicting fracture strength across the training, test, and validation datasets, demonstrating minimal prediction error across all datasets

5.4 Summary

This chapter established both the foundational and applied aspects of integrating a machine learning framework with molecular dynamics simulations to develop a high-fidelity predictive model for efficiently estimating the fracture strength of low-dimensional materials, specifically graphene. The chapter systematically discusses the various stages of the data-driven approach and their significance in building an effective ML pipeline. The predictive framework developed here is generic and can be extended to a wide range of material systems, provided a sufficiently large initial dataset is available. Additionally, this chapter offers practical insights into constructing an ML pipeline within the Python environment, making it a valuable reference for future applications in materials science and engineering.

5.5 How to Use the Source Files Associated with This Chapter?

The source files needed to reproduce the investigation laid out in this chapter are available in the GitHub repository linked at https://github.com/KriteshKG/ML4AtomisticSimulations/tree/main/Chapter_5

The link consists of two different folders one for conducting the MD simulations and the other folder for ML model formation.

5.5.1 MD Simulation

The folder pertaining to MD simulations (named MD_Simulations) consists of the required files to perform the nanoscale simulation of uniaxial tensile deformation over different configurations of monolayered graphene. The MD_Simulations folder namely consists of the following files:

LAMMPS input script:		*input_deformation.in*
Tersoff IP:		*SiC.tersoff*
Datafiles:	• Perfect graphene:	*data_perfect.xyz*
	• Single vacancy (SV) defected graphene:	*data_sv.xyz*
	• Si-doped graphene:	*data_si.xyz*
	• Combined SV and Si graphene:	*data_combined_sv_si.xyz*

At a time, the set of LAMMPS input scripts, Tersoff IP, and specific datafile can be used to perform the MD simulation. To run the simulation, the terminal is to be opened in the directory, and the following string of instructive code is to be entered [13]:

lmp -in input_deformation.in -pk omp 4 -sf omp

The same LAMMPS input script can be used by changing the specific command within the script:

read_data <use the specific file>.xyz

The MD simulations result in a text file which records the increment in strain and the corresponding stress in *X* (along the loading direction) and *Y* (lateral direction), which can be used to deduce the stress–strain profile of the graphene configurations, in terms of fracture strength.

5.5.2 Machine Learning

Another folder in the provided GitHub repository, named Machine Learning, consists of two files, first, the Excel file (Dataset.xlsx) which is the dataset obtained from the MD simulations, and, second, the Python (COLAB) notebook for executing the ML model-based investigation reported in this chapter. This notebook can be opened within the Google COLAB environment or by using Jupyter Notebooks on the local machine.

References

1. Siahlo AI, Popov AM, Poklonski NA, Lozovik YE, Vyrko SA (2020) Graphene membrane-based NEMS for study of interface interaction. Physica E: Low Dimen Syst Nanostruct 115:113645
2. Zhu J, Huang X, Song W (2021) Physical and chemical sensors on the basis of laser-induced graphene: mechanisms, applications, and perspectives. ACS Nano 15(12):18708–18741
3. Hou D, Zhang S, Chen X, Song R, Zhang D, Yao A, Sun J, Wang W, Sun L, Chen B, Liu Z (2021) Decimeter-scale atomically thin graphene membranes for gas-liquid separation. ACS Appl Mater Interfaces 13(8):10328–10335
4. Selvam SK, Kumar SRA, Balasubramanian N, Kumar PS, Muthukrishnaraj A, Kalaivani SS, Vinayagam R, Azarudeen RS, Ahamed MAR, Harikrishnan A, Josephine GS (2024) Recent advancements in the environmental significance of the graphitic carbon nitride-based materials. Desalin Water Treat:100385
5. Chandra Y, Adhikari S, Mukherjee S, Mukhopadhyay T (2022) Unfolding the mechanical properties of buckypaper composites: nano-to macro-scale coupled atomistic-continuum simulations. Eng Comp 38(6):5199–5229
6. Lee C, Wei X, Kysar JW, Hone J (2008) Measurement of the elastic properties and intrinsic strength of monolayer graphene. Science 321(5887):385–388
7. Gupta KK, Mukhopadhyay T, Roy A, Dey S (2020) Probing the compound effect of spatially varying intrinsic defects and doping on mechanical properties of hybrid graphene monolayers. J Mater Sci Technol 50:44–58
8. Gupta KK, Dey S (2019) Effect of temperature on the fracture strength of perfect and defective monolayered graphene. In: Advances in computational methods in manufacturing: select papers from ICCMM 2019. Springer Singapore, pp 793–804
9. Gupta KK, Roy A, Dey S (2020) Comparative study of various defects in monolayer graphene using molecular dynamics simulation. In: Advances in applied mechanical engineering: select proceedings of ICAMER 2019. Springer Singapore, pp 539–546
10. Gupta KK, Mukhopadhyay T, Dey S (2023) Probing the molecular-level energy absorption mechanism and strategic sequencing of graphene/Al composite laminates under high-velocity ballistic impact of nano-projectiles. Appl Surf Sci 629:156502
11. Gupta KK, Mukhopadhyay T, Roy A, Roy L, Dey S (2021) Sparse machine learning assisted deep computational insights on the mechanical properties of graphene with intrinsic defects and doping. J Phys Chem Solids 155:110111
12. Mahata A, Mukhopadhyay T, Adhikari S (2016) A polynomial chaos expansion based molecular dynamics study for probabilistic strength analysis of nano-twinned copper. Mater Res Exp 3(3):036501
13. LAMMPS, Running LAMMPS on Windows. Sandia National Laboratories. Available at: https://docs.lammps.org/Run_windows.html. Accessed 29 Apr 2025

Chapter 6
Nanoscale Ballistic Response of Bilayer Graphene: ML-Driven Approach

6.1 Overview

The investigation presented in the previous chapter focused on the quasi-static mechanical characterization of graphene layers, examining the fracture response of different graphene monolayer variants across a wide range of operating temperatures. To provide a more comprehensive understanding of graphene's mechanical behaviour, this chapter employs regression and classification machine learning (ML) models in conjunction with molecular dynamics (MD) simulations to characterize the transverse strength of bilayer graphene (BLG) configurations. The results explore the potential of BLG for nanoscale ballistic applications and reveal the damage patterns exhibited across a range of projectile impact velocities.

The low strength-to-weight ratio of conventional ballistic-proof materials has spurred significant interest in the development of novel materials that combine high strength with low weight. Among these, graphene stands out as a promising candidate due to its exceptional mechanical properties—it is the lightest known material (density $\approx$ 0.77 mg/m^2) and exhibits an extraordinary tensile strength of approximately 130 GPa. Recent research on graphene-reinforced composites has demonstrated that the incorporation of graphene imparts remarkable improvements in mechanical and ballistic performance [1–4]. A pioneering study by Lee et al. [5] employed a laser-induced projectile impact test on multilayer graphene (thickness $\approx$ 10–100 nm) and found that it could absorb kinetic energy from projectiles nearly ten times more effectively than macroscopic steel at a velocity of 600 m/s. Further investigations by Avila et al. [6] and O'Masta et al. [7] into graphene-reinforced composites confirmed a significant enhancement in ballistic resistance with the addition of graphene. These experimental insights have catalysed a growing research interest in the ballistic performance of graphene and other nanomaterials, leading to studies by Dewapriya and Meguid [8], Bizao et al. [9], Xia et al. [10], and Haque et al. [11], which continue to explore and validate the potential of nanostructured reinforcements in protective materials. The observations drawn from these experimental

K. K. Gupta et al., *Machine Learning in Nanoscale Materials Design*, Materials Horizons: From Nature to Nanomaterials, https://doi.org/10.1007/978-981-95-2660-4_6

studies have sparked considerable interest in exploring the ballistic capabilities of graphene and other nanomaterials. The advent of enhanced computational resources has further enabled researchers to adopt molecular dynamics (MD) simulations as a viable alternative to experimental characterization at the nanoscale. In recent years, the application of MD simulations to investigate high-velocity ballistic impacts on advanced materials has grown significantly within the research community [12–16].

This chapter utilizes the generalization capability of the ML models to completely characterize the ballistic performance of two critical configurations of BLG when initial impact velocity is varied from 1 to 7 km/sec. This range of velocities encompasses typical projectile speeds observed in military combat scenarios as well as those encountered in space exploration and aerospace applications. In this framework, the projectile's impact velocity (V_i) along with the BLG configuration (AA or AB) serves as the input features, while the specific penetration energy (E_p*) of the BLG is treated as the output variable for the regression task. In addition to the quantitative findings, the qualitatively analysed post-impact behaviour of BLG categorized it into four distinct stages—*R* (rebound of the projectile without damage to BLG), PP1 (rebound of the projectile with partial penetration when first layer got damaged, and second layer remains intact), PP2 (rebound of the projectile with partial penetration when both layers got damaged), and CP (complete penetration of the projectile)—based on the level of damage and the projectile's post-impact trajectory, enabling a classification task.

6.2 MD Simulation-Driven Dataset

In this chapter, two types of bilayer graphene (BLG) configurations—AA stacking (where two graphene layers are perfectly aligned with an interlayer spacing of 3.4 Å) and AB stacking (where one layer is laterally shifted with the same interlayer distance)—are modelled as square sheets of 200 Å × 200 Å. These BLG structures are subjected to high-velocity impact from a spherical diamond projectile with a diameter of approximately 25 Å (refer to Fig. 6.1). The interatomic interactions between the carbon atoms present in the individual graphene layers are modelled in AIREBO-Morse [17] interatomic potential. The interface between two monolayers of graphene is modelled by Van der Waals force. The projectile is described using the Tersoff [18] potential, which is widely recognized as an appropriate force field for capturing carbon–carbon (C–C) interactions in diamond. The initial distance between the centre of the projectile and the top graphene layer is set to 75 Å to allow for acceleration before impact. The interactions between the BLG and the diamond projectile are modelled using the Morse potential, which effectively captures non-bonded interactions between the two materials.

The MD simulations of high-velocity impact are carried out by applying periodic boundary conditions along the in-plane directions (x and y) of the bilayer graphene (BLG). To simulate clamped edges, a 5 Å-wide region along all four sides of the BLG sheet is fixed by setting both force and velocity components to zero. A small

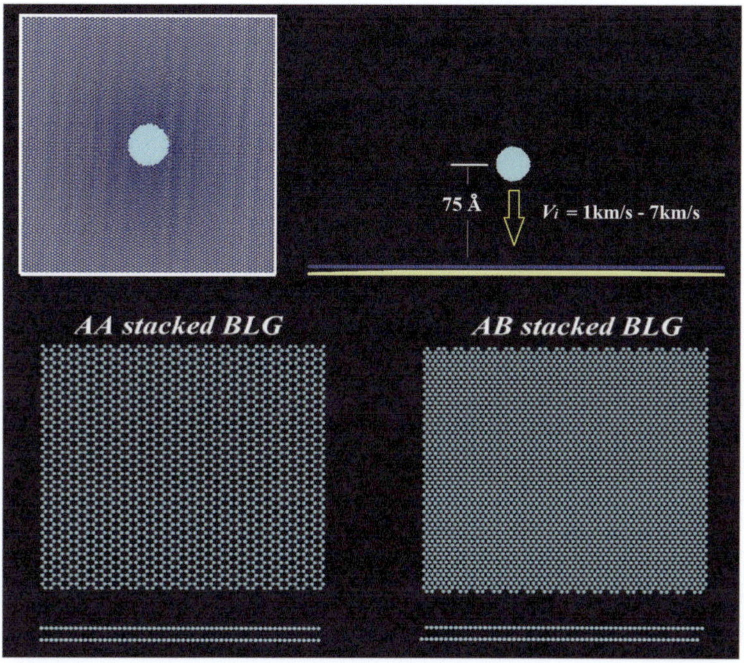

Fig. 6.1 Atomistic model of bilayer graphene (BLG configurations utilized in the investigation for high-velocity impact)

time step of 0.5 femtoseconds (fs) is used throughout the simulations to ensure numerical stability. The simulation process begins with a 25 picosecond (ps) energy minimization using the conjugate gradient method to relax the atomic configuration and eliminate residual stresses [19, 20]. This is followed by a 25 ps thermal equilibration phase, during which the system is brought to room temperature (~ 300 K) using the Berendsen thermostat. Finally, the high-velocity impact simulation is conducted under a microcanonical (NVE) ensemble for an additional 20 ps, during which the rigid diamond projectile is launched towards the BLG at a predefined velocity.

The ballistic performance of the BLG structure is assessed on the basis of the specific penetration energy ($E_p{}^*$) of the BLG, and the behaviour of BLG against the high-velocity impact. The specific penetration energy of the BLG is determined by using the following relation [21–24]:

$$E_{\mathrm{p}}^{*} = \frac{\mathrm{KE_i} - \mathrm{KE_r}}{M} \tag{6.1}$$

The KE_i and KE_r refer to the initial and residual kinetic energy of the projectile, respectively. The difference of both kinetic energies is divided by the term M which denotes the mass of the impacted zone. The mass of the impacted zone is determined by the following equation:

$$M = \pi . R^2 . N_l . \rho_A \quad (6.2)$$

where R denotes the radius of the impacted zone (same as a projectile), N_l denotes the number of layers, and ρ_A denotes the area density of graphene which is taken as $0.77e^{-6}$ kg/m^2 in the present study [11].

6.3 Machine Learning Pipeline

The dataset used in this study comprises 320 samples, with 160 samples each corresponding to AA- and AB-stacked bilayer graphene (BLG) configurations (refer to Fig. 6.1). It is generated through a series of molecular dynamics (MD) simulations involving high-velocity impacts, where the impact velocity is continuously varied between 1 and 7 km/s. Accordingly, the dataset used for machine learning (ML) model development includes two input features: the BLG stacking configuration (a categorical variable) and the impact velocity (a continuous variable). The specific penetration energy ($E_p{}^*$) is considered the response variable for the regression model.

For the classification task, the post-impact behaviour of the BLG is categorized into four distinct damage stages:

- *R*: Rebound of the projectile without any damage to the BLG;
- PP1: Rebound of the projectile with partial penetration, where the first graphene layer is damaged but the second remains intact;
- PP2: Rebound with partial penetration, where both graphene layers are damaged;
- CP: Complete penetration of the projectile through the BLG.

The primary objective of the regression model is to build a predictive framework capable of estimating the ballistic response—specifically, the penetration energy—of BLG structures subjected to previously unseen impact velocities. In parallel, the classification model aims to develop a predictive pipeline that can accurately determine the extent of structural damage to BLG configurations under high-velocity impact conditions.

As discussed in Sect. 5.3 of Chap. 5, the ML pipeline in this chapter also starts with the preprocessing of the input features, where one-hot encoding is used to numerically represent the BLG stacking configurations, while the impact velocity is preprocessed by using the z-score normalization.

In the following stage, the best possible regression and classification ML models are selected by using the *LazyPredict* regressor, and *LazyPredict* classifier. As a result, a gradient boosting regressor and light gradient boosting machine (LGBM) classifier are selected to develop the ML models. Similar to the ML pipeline discussed in Chap. 5, the regression and classification models are trained by enforcing a k-fold ($k = 5$) cross-validation scheme during hyperparameter tuning.

The consistency of the GBR ML model is assessed by observing the R^2 values across the 5 folds of k-fold cross-validation (refer to Fig. 6.2).

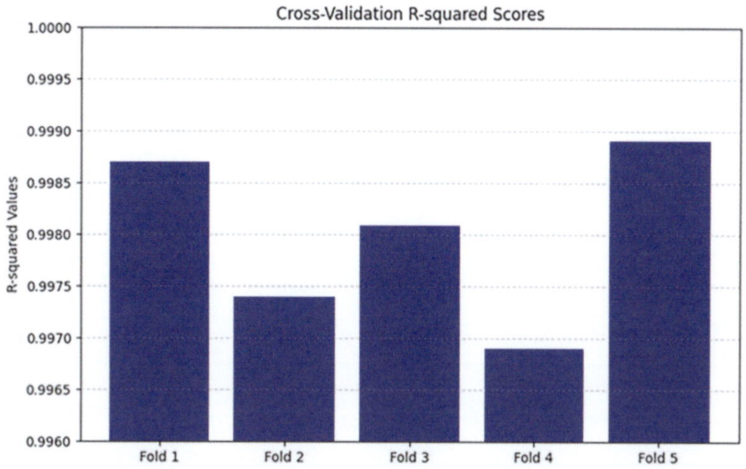

Fig. 6.2 Observed R^2 values across all the folds of fivefold cross-validation

The validation of the Gradient Boosting Regression (GBR) model is presented in Fig. 6.3, where the GBR-based predictions are compared against the specific penetration energy values obtained from MD simulations.

The predictive accuracy is evaluated across the training, test, and validation datasets by analysing the probability density function (PDF) of the sample-wise relative error. The results indicate that, for the majority of the data samples, the relative prediction error falls within the range of $\pm$ 0.2, demonstrating the model's

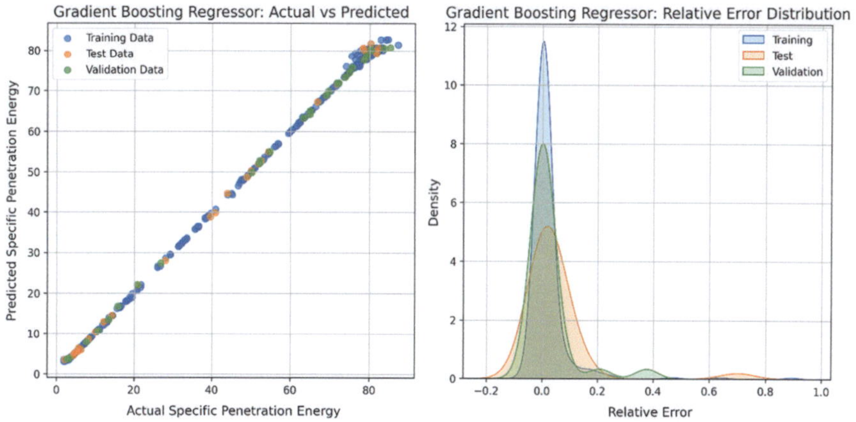

Fig. 6.3 ML validation for gradient boosting regressor by using the scatter plots for training, test, and validation datasets (demonstrates how close the ML-predicted specific penetration energy is with the MD simulation-based specific penetration energy) and probability density function (pdf) plots for the distribution of relative error (error w.r.t. actual specific penetration energy) for training, test, and validation datasets

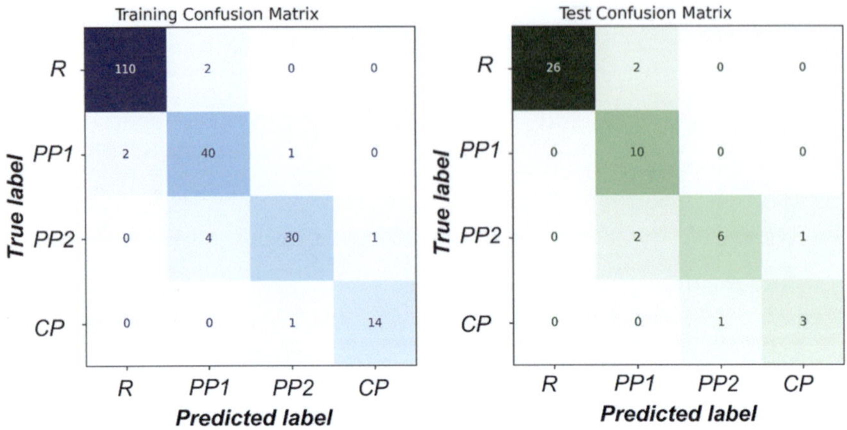

Fig. 6.4 Confusion matrix for LGBM classifier demonstrating the extent of misclassification made by the model for training and test datasets

strong predictive performance. The developed regression model exhibits exceptional predictive accuracy, even for the unseen dataset (which was not exposed to the model during model construction).

Next, the classification dataset (consisting of BLG configurations and impact velocity as the input features and extent of damage as classification labels) is used to develop the LGBM classification model.

Such a classification model will be capable of classifying the extent of damage in BLG which is induced by high-velocity impact. The confusion matrices for the training and test datasets are illustrated in Fig. 6.4, which reveals the classification accuracy of the developed classification model. It is evident that regardless of the training and test datasets, the classification model demonstrates significant accuracy across the class labels, with minimal classifications.

The performance measures for the classification models, such as precision, recall, and $F1$-score, across the class labels for training and test datasets are illustrated in Fig. 6.5, which reveals the exceptional classification performance of the trained LGBM classifier.

6.4 Summary

In summary, this chapter showcases the successful application of machine learning-based regression and classification frameworks for thoroughly characterizing the nanoscale ballistic performance of bilayer graphene. This chapter integrated a set of ensemble regression and classification models with the dataset derived from a series of MD simulations. In this regard, the considered input parameter (impact velocity (V_i)) is varied from 1 (10 Å/ps) to 7 km/s (70 Å/ps) using Monte Carlo sampling,

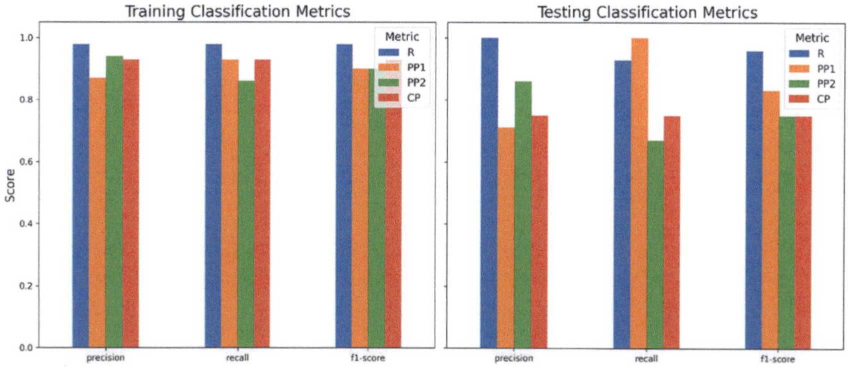

Fig. 6.5 Variation in performance measures of LGBM classifier across all the considered classification labels for training and test datasets

generating a dataset of 128 simulation instances for two different configurations of BLG, namely AA and AB. Each simulation involved a spherical diamond projectile (25 Å in diameter) impacting a BLG sheet (200 Å × 200 Å) at high velocity. For each MD simulation, the specific penetration energy (E_p^*) is recorded as the output response. In addition to the quantitative findings, the qualitatively analysed post-impact behaviour of BLG categorized it into four distinct stages—*R* (rebound of the projectile without damage to BLG), PP1 (rebound of the projectile with partial penetration when first layer got damaged, and second layer remains intact), PP2 (rebound of the projectile with partial penetration when both layers got damaged), and CP (complete penetration of the projectile)—based on the level of damage and the projectile's post-impact trajectory. The dataset generated through a series of MD simulations is then used to train ensemble ML models for both regression and classification tasks. Through this methodology, the chapter showcases the integrated framework of MD simulations and ML (regression and classification) to accelerate the nanoscale material characterization by combining the predictive power of ML with the accuracy of MD simulations. It presents the data-driven approach chronologically, highlighting the role of each stage in constructing an effective ML pipeline. The ML-enabled quantitative and qualitative analyses provide valuable insights into damage patterns and the impact resistance of bilayer graphene as influenced by varying impact velocities.

6.5 How to Use the Source Files Associated with This Chapter?

The source files needed to reproduce the investigation laid out in this chapter are available in the GitHub repository linked at https://github.com/KriteshKG/ML4AtomisticSimulations/tree/main/Chapter_6

The link consists of two different folders, one for conducting the MD simulations and the other folder for ML model formation.

6.5.1 MD Simulation

The folder pertaining to MD simulations (named MD_Simulations) consists of the required files to perform the nanoscale simulation of high-velocity impact over bilayer graphene. The MD_Simulations folder namely consists of the following files:

LAMMPS input script:	*input.in*
Tersoff IP: (for modelling silicon impactor)	*SiC.tersoff*
AIREBO-m: (for modelling bilayer graphene)	*CH.airebo-m*
Datafiles:	*BLG_20_20.xyz*

The set of LAMMPS input scripts, IP files, and the datafile can be used to perform the MD simulation. To run the simulation, the terminal is to be opened in the directory, and the following string of instructive code is to be entered:

lmp -in input.in -pk omp 4 -sf omp

The MD simulations result in a text file named 'output_data_bball.txt' which records the temporal variation in kinetic energy of the impactor, which is used to obtain the pre-impact and post-impact kinetic energy of the projectile, which is then used to calculate specific penetration energy by utilizing Eq. 6.1.

6.5.2 Machine Learning

Another folder in the provided GitHub repository, named Machine Learning consists of three files, first, the Excel file (Data_regression.xlsx) which is the dataset for the discussed regression problem, second, the Excel file (Data_classification.xlsx) which is the dataset for the discussed classification problem, and the third file (Chapter_6.ipynb) is a Python (COLAB) notebook for executing the ML model-based investigation reported in this chapter. This notebook can be opened within the Google COLAB environment or by using Jupyter Notebooks on the local machine.

References

1. Stankovich S, Dikin DA, Dommett GH, Kohlhaas KM, Zimney EJ, Stach EA, Piner RD, Nguyen ST, Ruoff RS (2006) Graphene-based composite materials. Nature 442(7100):282–286

2. Kwon SR, Harris J, Zhou T, Loufakis D, Boyd JG, Lutkenhaus JL (2017) Mechanically strong graphene/aramid nanofiber composite electrodes for structural energy and power. ACS Nano 11(7):6682–6690
3. Li Z, He Q, Xu X, Zhao Y, Liu X, Zhou C, Ai D, Xia L, Mai L (2018) A 3D nitrogen-doped graphene/TiN nanowires composite as a strong polysulfide anchor for lithium-sulfur batteries with enhanced rate performance and high areal capacity. Adv Mater 30(45):1804089
4. Shingare KB, Kundalwal SI (2019) Static and dynamic response of graphene nanocomposite plates with flexoelectric effect. Mech Mater 134:69–84
5. Lee JH, Loya PE, Lou J, Thomas EL (2014) Dynamic mechanical behavior of multilayer graphene via supersonic projectile penetration. Science 346(6213):1092–1096
6. Ávila AF, Neto AS, Junior HN (2011) Hybrid nanocomposites for mid-range ballistic protection. Int J Impact Eng 38(8–9):669–676
7. O'Masta MR, Russell BP, Deshpande VS (2017) An exploration of the ballistic resistance of multilayer graphene polymer composites. Extreme Mech Lett 11:49–58
8. Dewapriya MA, Meguid SA (2019) Comprehensive molecular dynamics studies of the ballistic resistance of multilayer graphene-polymer composite. Comput Mater Sci 170:109171
9. Bizao RA, Machado LD, de Sousa JM, Pugno NM, Galvao DS (2018) Scale effects on the ballistic penetration of graphene sheets. Sci Rep 8(1):1–8
10. Xia K, Zhan H, Hu DA, Gu Y (2016) Failure mechanism of monolayer graphene under hypervelocity impact of spherical projectile. Sci Rep 6(1):1
11. Haque BZ, Chowdhury SC, Gillespie JW Jr (2016) Molecular simulations of stress wave propagation and perforation of graphene sheets under transverse impact. Carbon 102:126–140
12. Dewapriya MA, Miller RE (2021) Energy absorption mechanisms of nanoscopic multilayer structures under ballistic impact loading. Comput Mater Sci 195:110504
13. Dewapriya MA, Miller RE (2020) Molecular dynamics study of the mechanical behaviour of ultrathin polymer–metal multilayers under extreme dynamic conditions. Comput Mater Sci 184:109951
14. Dewapriya MA, Miller RE (2021) Molecular dynamics study of the penetration resistance of multilayer polymer/ceramic nanocomposites under supersonic projectile impacts. Extreme Mech Lett 44:101238
15. Dewapriya MA, Miller RE (2020) Superior dynamic penetration resistance of nanoscale multilayer polymer/metal films. J Appl Mech 87(12):121009
16. Chandra Y, Mukhopadhyay T, Adhikari S (2020) Size-dependent dynamic characteristics of graphene based multi-layer nano hetero-structures. Nanotechnology 31(14):145705
17. O'Connor TC, Andzelm J, Robbins MO (2015) AIREBO-M: a reactive model for hydrocarbons at extreme pressures. J Chem Phys 142(2):024903
18. Tersoff J (1988) New empirical approach for the structure and energy of covalent systems. Phys Rev B 37(12):6991
19. Gupta KK, Roy A, Dey S (2020) Comparative study of various defects in monolayer graphene using molecular dynamics simulation. In: Advances in applied mechanical engineering: select proceedings of ICAMER 2019. Springer Singapore, pp 539–546
20. Gupta KK, Dey S (2019) Effect of temperature on the fracture strength of perfect and defective monolayered graphene. In: Advances in computational methods in manufacturing: select papers from ICCMM 2019. Springer Singapore, pp 793–804
21. Gupta KK, Roy L, Dey S (2022) Ballistic performance of bi-layer graphene: artificial neural network based molecular dynamics simulations. In: Machine learning applied to composite materials. Springer Nature Singapore, Singapore, pp 135–148
22. Gupta KK, Mukhopadhyay T, Dey S (2023) Probing the molecular-level energy absorption mechanism and strategic sequencing of graphene/Al composite laminates under high-velocity ballistic impact of nano-projectiles. Appl Surf Sci 629:156502
23. Gupta KK, Mukhopadhyay T, Roy L, Dey S (2022) High-velocity ballistics of twisted bilayer graphene under stochastic disorder. Adv Nano Res 12(5):529–547
24. Gupta KK, Barman S, Dey S, Naskar S, Mukhopadhyay T (2024) On exploiting nonparametric kernel-based probabilistic machine learning over the large compositional space of high entropy alloys for optimal nanoscale ballistics. Sci Rep 14(1):16795

Chapter 7
Interatomic Potential Parametrization for Graphene

7.1 Overview

In Chaps. 5 *and* 6, *different configurations of graphene were designed and characterized for fracture strength and impact strength, respectively, by using the molecular dynamics (MD) simulation-driven ML model. MD simulations have become an essential tool for studying nanoscale materials such as graphene. However, the accuracy of the results from MD simulations relies heavily on the interatomic potential (IP) used, which is mathematically calibrated based on data from first-principles methods or experiments. There is considerable uncertainty associated with the parameters of the IP. Such internal uncertainties can trigger an augmented random deviation in the mechanical responses. Hence, this chapter utilizes an explainable ML model for the parametrization of Tersoff IP parameters to model the fracture response of graphene close to the experimentally reported results.*

The previous literature suggests that the utilization of the original Tersoff interatomic potential (IP) to simulate the uniaxial tensile deformation of graphene leads to the overestimation of fracture strength and induces the unphysical strain hardening in the graphene structure before fracture. It is reported by Lindsay and Broido [1] that the original parameters of Tersoff IP have room for optimizing certain parameters to exactly produce the phonon dispersion of graphene. Further, Mortazavi et al. [2] reported that the exact experimental results of the fracture strength of graphene can be produced by modifying the cut-off distance from 1.95 to 2.0 angstrom. Rajsekaran et al. [3] reproduced the exact experimental results of fracture characteristics of graphene by utilizing changes suggested by Lindsay and Broido [1] and Mortazavi et al. [2]. This indicates that the values of the parameter set of the Tersoff IP are not constant, and their random perturbation can significantly influence the MD simulation results [4]. Hence it leads to a strong rationale to parametrize the Tersoff IP parameters to characterize the accurate fracture response of graphene. To mitigate this lacuna, the emerging capabilities of machine learning are exploited, wherein the

K. K. Gupta et al., *Machine Learning in Nanoscale Materials Design*, Materials Horizons: From Nature to Nanomaterials, https://doi.org/10.1007/978-981-95-2660-4_7

expensive MD simulations can effectively be replaced by an efficient mathematical model.

In the present chapter, the MD simulation-based deterministic analysis of graphene has been extended to the stochastic regime for carrying out a comprehensive analysis pertaining to the variations in Tersoff IP parameters.

7.2 Machine Learning-Driven Feature Importance of Tersoff IP

In this section, the explainable machine learning (ML) assisted computational mapping between the uncertainties associated with the Tersoff IP, and the fracture response of graphene is established. To form the ML model, random sampling of the parameters of Tersoff IP is carried out with 2% variations to create the design space of $N = 128$ samples. This is carried out by considering the mean values of Tersoff IP parameters (given in Table 7.1) and the perturbation range of 2% to generate 128 random samples. The randomly sampled Tersoff IP parameters are enforced while performing the MD simulations of the uniaxial tensile deformation of graphene and recording the fracture strength of graphene. The dataset is utilized to develop the ML model by using the GPR pipeline illustrated in Fig. 7.1.

The validation of the developed GPR model is performed by using the scatter plots and error plots (illustrated in Fig. 7.2), which reveals that the ML-driven prediction of fracture strength of graphene remains very close to the actual fracture strength (MD simulation-driven).

Table 7.1 Internal (Tersoff IP) parameters of MD simulation considered as sources of uncertainty [5]

Type of uncertainty source	Parameters	Deterministic values
Internal parameters (Tersoff)	□	1.0
	A (eV)	1393.6
	B (eV)	346.74
	λ_1 (Å^{-1})	3.4879
	λ_2 (Å^{-1})	2.2119
	N	0.72751
	C	38,049
	β ($\times 10^{-7}$)	1.5724
	D	4.3484
	H	− 0.57058
	R (Å)	1.8
	D (Å)	0.15

```
# Define a function to train and tune a GPR model
def train_gpr(X_train, Y_train):
    # Define a pipeline with feature scaling and GPR
    pipeline = Pipeline([
        ('scaler', StandardScaler()),
        ('gpr', GaussianProcessRegressor())
    ])

    # Define the hyperparameter grid
    param_grid = {
        'gpr__kernel': [
            C(1.0) * RBF(length_scale) for length_scale in [0.1, 1.0, 10.0]
        ],
        'gpr__alpha': [1e-10, 1e-5, 1e-2]
    }

    # Perform grid search with cross-validation
    grid_search = GridSearchCV(pipeline, param_grid, cv=5,
                               scoring='neg_mean_squared_error', n_jobs=-1)
    grid_search.fit(X_train, Y_train)

    return grid_search.best_estimator_, grid_search.best_params_
```

Fig. 7.1 The Gaussian process regression pipeline designed for developing a hyperparameter-tuned GPR model

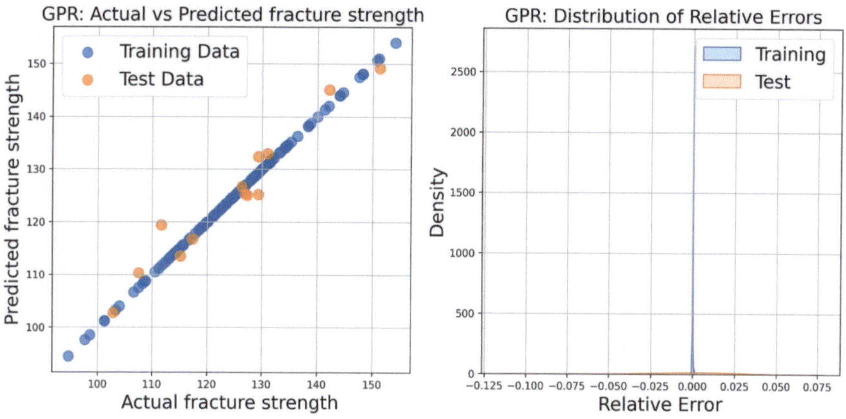

Fig. 7.2 Validation of GPR model for training and test dataset by using scatter plot, and pdf of relative error

Once, the effective prediction capabilities of the GPR model are established, the model is deployed for making large-scale predictions. To perform this, the virtual sample space ($\approx$ 10,000 samples) of perturbed Tersoff IP is generated by using random sampling, by enforcing 4 different perturbation configurations. Such perturbation aims to intentionally induce the uncertainty in the Tersoff IP parameters and obtain the influence of such uncertainty on the fracture response of graphene. In this regard, perturbations are performed with the magnitude of 0.5%, 1.0%, 1.5%, and 2.0%. The perturbed Tersoff IP parameters are fed to the GPR model to make

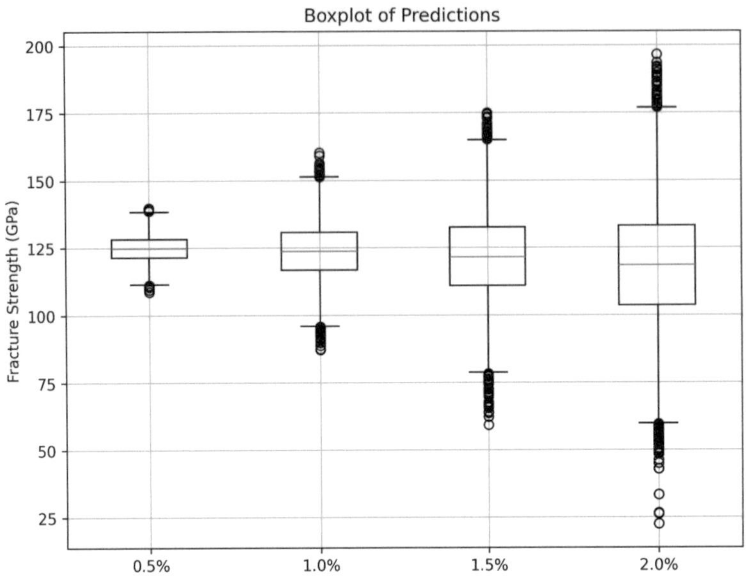

Fig. 7.3 Influence of uncertainty (0.5%, 1.0%, 1.5%, and 2.0%) in Tersoff IP parameters on the MD simulation-derived fracture strength of graphene

large-scale predictions of fracture strength. The observations obtained from the large-scale predictions are illustrated in Fig. 7.3 as boxplots to demonstrate the influence of uncertainty in Tersoff IP parameters on the MD simulation response.

As a primary outcome, it is revealed from Fig. 7.3 that the variation in the range of fracture strength gradually increases as the variation in the uncertainty perturbation of the Tersoff IP parameters is increased (from 0.5 to 2.0%). To compare the obtained ML-driven observations with the reported Tersoff IP modifications in the past, the stress–strain profile of graphene monolayer obtained by using the original Tersoff IP is compared with the profiles obtained by using modified Tersoff proposed by Mortazavi et al. [2] and optimized Tersoff proposed by Rajasekaran et al. [3] (refer to Fig. 7.4).

It is evident from the stress–strain profiles presented in Fig. 7.4 that by using a modified and optimized Tersoff force field, the unphysical strain hardening is eliminated but at the same time the failure strain (experimental value ≈ 0.25) is greatly compromised. It is evident that the parametric uncertainty in Tersoff IP greater than 0.5% results in the fracture strength of graphene which is far away from the reported experimental observation of 130 ± 10 GPa [6] (as illustrated in Fig. 7.4). Hence, the parametric perturbations of 0.5% in Tersoff IP are more suitable for providing the accurate experimental fracture response of graphene. With this understanding, at first, the feature importance (out of considered 12 features) has to be carried out, to showcase the influence of individual features on the prediction of fracture strength of graphene. To perform the interpretation of the developed ML models and associated prediction-making mechanism, SHapley Additive exPlanations (SHAP) is deployed

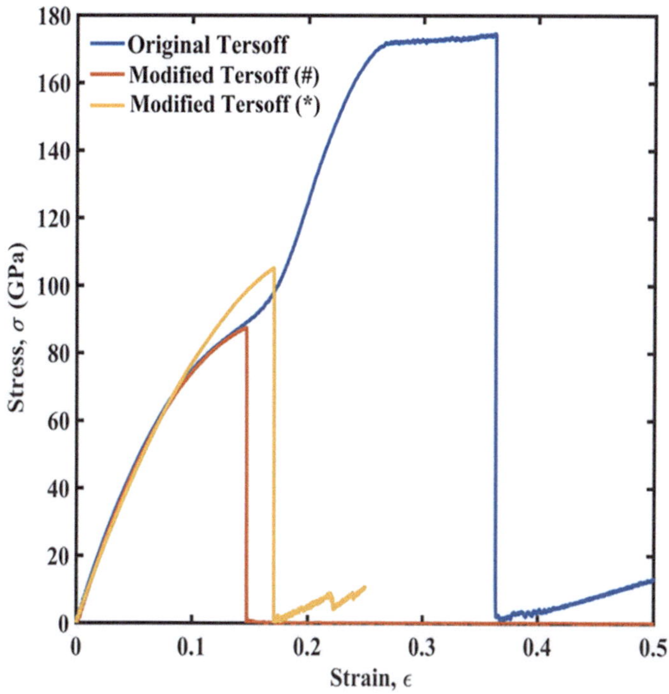

Fig. 7.4 Comparison of stress–strain performance for original Tersoff, and modified Tersoff (#Mortazavi et al. [2], *Rajasekaran et al. [3])

to reveal the contribution of an individual parameter on the prediction of fracture response of graphene. The Python code snippet for SHAP-based explanation of the trained GPR model is provided in Fig. 7.5.

```
!pip install shap
import shap

# Extract scaler and model
scaler = best_gpr.named_steps['scaler']
gpr_model = best_gpr.named_steps['gpr']
X_scaled = scaler.transform(X)

# Create SHAP KernelExplainer with 100-sample background
background = shap.utils.sample(X_scaled, 100, random_state=42)
explainer = shap.KernelExplainer(gpr_model.predict, background)

# Calculate SHAP values for entire dataset
shap_values = explainer.shap_values(X_scaled)
```

Fig. 7.5 Python code snippet to perform explanation of GPR models by obtaining the SHAP values

The obtained SHAP observations are plotted in terms of bar plot and bee-swarm plot (Python code snippet provided in Fig. 7.6) to reveal the feature importance of the individual Tersoff IP parameter in predicting the fracture response. Such feature-specific information helps in understanding the role of individual features in the prediction-making mechanism of the ML model, and how the feature variation will affect the target response (fracture strength in our case). The bar plot illustrated in Fig. 7.7a highlights the relative importance of the Tersoff IP parameters, which reveals that the terms λ_1 and λ_2 are having the highest influence in predicting the fracture response of graphene. Hence, the variations in these two terms can have a significant impact on the accuracy of the predicted responses. The SHAP summary illustrated in Fig. 7.7b as a bee-swarm plot reveals that the higher value of λ_1 results in enhanced fracture strength (since + ve SHAP values are obtained), and vice versa. In contrast, the higher value of λ_2 leads to a decreased value of fracture strength of graphene (since -ve SHAP values are obtained), and vice versa.

```python
# Plot summary SHAP bar chart
fig, ax = plt.subplots(figsize=(12, 8))
shap.summary_plot(shap_values, X, plot_type="bar", show=False)
plt.title("SHAP Summary Plot (Bar)", fontsize=20, fontweight='bold')
plt.xlabel("Mean Absolute SHAP Value", fontsize=16)
plt.ylabel("Feature", fontsize=16)
plt.savefig("shap_summary_bar.png", bbox_inches='tight')
plt.show()

# Detailed SHAP value distribution
fig, ax = plt.subplots(figsize=(12, 8))
shap.summary_plot(shap_values, X, show=False)
plt.title("SHAP Summary Plot (Beeswarm)", fontsize=20, fontweight='bold')
plt.xlabel("SHAP Value (Impact on Model Output)", fontsize=16)
plt.ylabel("Feature", fontsize=16)
plt.savefig("shap_summary_beeswarm.png", bbox_inches='tight')
plt.show()
```

Fig. 7.6 Python code snippet to plot the SHAP-based explanations

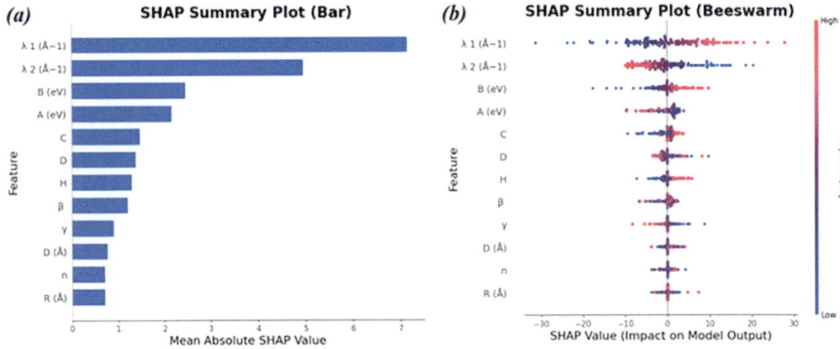

Fig. 7.7 SHAP-based explanation of GPR model: **a** relative feature importance; **b** SHAP summary plot

The observations obtained from the SHAP-based explanation provide significant information about the parametrization of Tersoff IP parameters for accurate fracture response of graphene. It is to be noted that the original parameters of Tersoff result in the overestimated fracture strength of graphene ($\approx$ 190 GPa), with considerable (unphysical) plastic deformation (refer to Fig. 7.4), whereas the experimental characterization of graphene results in the fracture strength of 130 $\pm$ 10 GPa [6]. To address this in the following section, the observations obtained from SHAP-based explanations are used to modify the Tersoff IP parameters.

7.3 Modification of Tersoff IP Parameters

It is revealed from the SHAP analysis presented in the preceding section that out of considered twelve Tersoff IP parameters, λ_1 and λ_2 are relatively more significant in determining the fracture strength of graphene.

Hence, in the following analysis, different combinations of λ_1 and λ_2 are utilized to perform the MD simulations and obtain the fracture behaviour of graphene (by keeping the rest of the parameters at their original values). The variation $\pm$ 0.5% is induced in the original values of λ_1 and λ_2 for this analysis. At first, the maximum and minimum values of λ_1 are used to compare the fracture response of graphene with the original Tersoff IP (refer to Fig. 7.8). It can be noticed from Fig. 7.8 that utilizing the maximum and minimum values of λ_1 reduces the unphysical strain hardening in the stress–strain response of the graphene up to a great extent. In addition to that, the maximum value of λ_1 leads to the fracture strength which is close to the experimental observation by Lee et al. [6]. The same practice is carried out for the variation in λ_2, which is illustrated in Fig. 8.8 of Chap. 8.

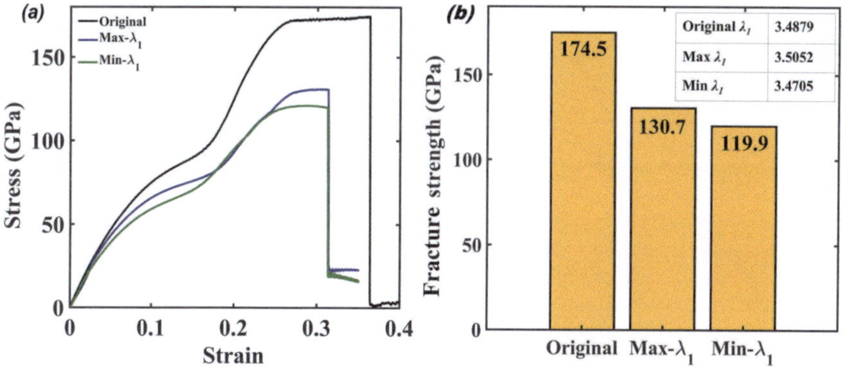

Fig. 7.8 Variation in fracture response of graphene with respect to the variation in λ_1: **a** Stress–strain behaviour of graphene for different λ_1 values; **b** fracture strength of graphene for different λ_1 values

Similar to the observations in Fig. 7.8, the observations from Fig. 7.9 also revealed that the minimum value of λ_2 results in the fracture strength of graphene which is close to the experimental observation by Lee et al. [6]. Hence, next, the combined variation in λ_1 and λ_2 is induced to compare the fracture behaviour of graphene under each case (refer to Fig. 7.10). It is evident from Fig. 7.10 that the maximum value of λ_1 and minimum value of λ_2 altogether results in a fracture strength close to the experimental observation reported by Lee et al. [6] (a similar observation was given by SHAP-based summary plot illustrated in Fig. 7.7b). It is worth mentioning that the proposed changes in the values of λ produce the accurate fracture strength of graphene (close to experimental results), by preventing the overestimation and compromise in failure strain (refer to Fig. 7.10).

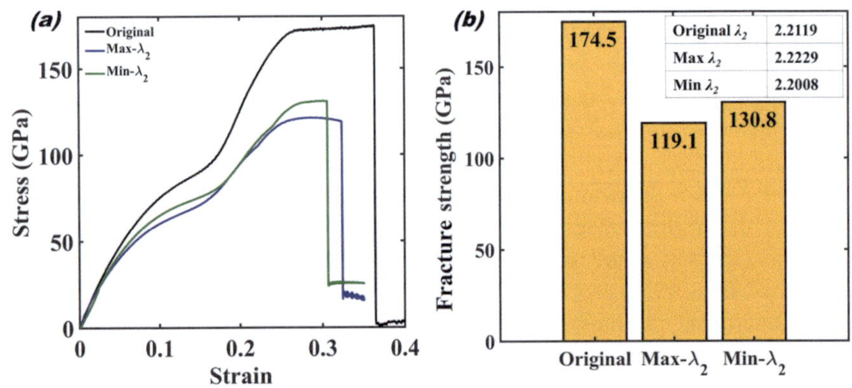

Fig. 7.9 Variation in fracture response of graphene with respect to the variation in λ_2: **a** Stress–strain behaviour of graphene for different λ_2 values; **b** fracture strength of graphene for different λ_2 values

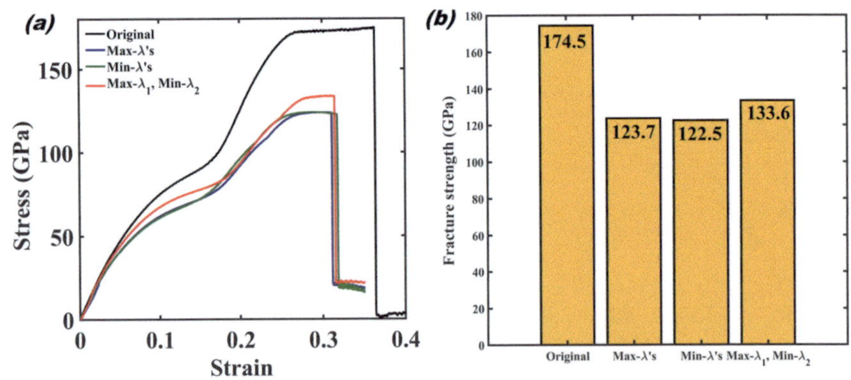

Fig. 7.10 Variation in fracture response of graphene with respect to the variation in both λ's: **a** Stress–strain behaviour of graphene for different values of λ; **b** fracture strength of graphene for different values of λ

7.4 Summary

In this chapter, a statistical analysis is carried out systematically to propose suitable changes in the Tersoff interatomic potential (IP) parameters which lead to accurate fracture response of graphene. The Tersoff potential is extensively utilized to perform MD simulations of graphene, where results significantly depend on the IP used for the specific analysis. These IPs are calibrated with the experimental data or the data received from ab initio methods to mimic certain properties of the system under consideration. When the simulations are conducted for the observations beyond the calibrated envelope of these IPs, the validity of predicted results is questionable. In general, there is always a degree of uncertainty associated with the values of interatomic potentials, which could significantly influence the outcome of MD simulations. In this regard, this chapter proposes the framework for performing uncertainty quantification of Tersoff IP parameters by considering the fracture strength of graphene as the quantity of interest. Herein, the random sampling (Sobol sequence) and Gaussian process regression-based machine learning are integrated with the molecular dynamics simulation to conduct the SHAP-based explanations. The study reveals that among the interatomic parameters, λ_1 and λ_2 are found to be relatively most sensitive. As an outcome, the present chapter proposes suitable changes in the values of λ_1 and λ_2 for obtaining the accurate fracture responses of graphene by using the modified Tersoff IP.

7.5 How to Use the Source Files Associated with This Chapter?

The source files needed to reproduce the investigation laid out in this chapter are available in the GitHub repository linked at https://github.com/KriteshKG/ML4AtomisticSimulations/tree/main/Chapter_7

The link consists of two different folders, one for conducting the MD simulations and the other folder for ML model formation.

7.5.1 *MD Simulation*

The folder pertaining to MD simulations (named MD_Simulations) consists of the required files to perform the nanoscale simulation of high-velocity impact over bilayer graphene. The MD_Simulations folder namely consists of the following files:

LAMMPS input script:	*input_1.in*
Tersoff IP:	*SiC.tersoff*

(continued)

(continued)

Datafiles:	*data_perfect.xyz*

The set of LAMMPS input scripts, IP files, and the datafile can be used to perform the MD simulation. To run the simulation, the terminal is to be opened in the directory, and the following string of instructive code is to be entered:

lmp -in input.in -pk omp 4 -sf omp

The MD simulations result in a text file which records the increment in strain and the corresponding stress in X (along the loading direction) and Y (lateral direction), which can be used to deduce the stress–strain profile of the graphene configurations, in terms of fracture strength.

7.5.2 *Machine Learning*

Another folder in the provided GitHub repository, named Machine Learning consists of two files, first, the Excel file (Dataset.xlsx) which is the dataset for the discussed regression problem, and, second, the Python notebook (Chapter_7.ipynb) for executing the ML model-based investigation reported in this chapter. This notebook can be opened within the Google COLAB environment or by using Jupyter Notebooks on the local machine. Other necessary Excel files used in the Python notebook are also provided in this folder.

References

1. Lindsay L, Broido DA (2010) Optimized Tersoff and Brenner empirical potential parameters for lattice dynamics and phonon thermal transport in carbon nanotubes and graphene. Phys Rev B 81(20):205441
2. Mortazavi B, Fan Z, Pereira LFC, Harju A, Rabczuk T (2016) Amorphized graphene: a stiff material with low thermal conductivity. Carbon 103:318–326
3. Rajasekaran G, Kumar R, Parashar A (2016) Tersoff potential with improved accuracy for simulating graphene in molecular dynamics environment. Mater Res Exp 3(3):035011
4. Trinh MC, Mukhopadhyay T (2021) Semi-analytical atomic-level uncertainty quantification for the elastic properties of 2D materials. Materials Today Nano 15:100126
5. Gupta KK, Mukhopadhyay T, Roy L, Dey S (2022) Hybrid machine-learning-assisted quantification of the compound internal and external uncertainties of graphene: towards inclusive analysis and design. Mater Adv 3(2):1160–1181
6. Lee C, Wei X, Kysar JW, Hone J (2008) Measurement of the elastic properties and intrinsic strength of monolayer graphene. Science 321(5887):385–388

Chapter 8
High-Entropy Alloy: A Data-Driven Quasi-Static Characterization

8.1 Overview

Chapters 5, 6, and 7 reported the ML-driven mechanical characterization of different graphene configurations, which are the lower dimensional nanomaterials. To extend a similar approach in the nanoscale design of high-dimensional materials systems, this chapter presents the ML-driven strategy for designing lightweight AlCoCr-FeNi high entropy alloys while enhancing Young's modulus. Designing high-entropy alloys (HEAs) is a challenging task due to their broad compositional space and the composition-dependent variation in their functional properties. Traditional alloy design typically involves trial-and-error prototyping and high-throughput experiments, which can be difficult because of the complexities involved in large-scale synthesis and characterization. To tackle these issues, this chapter presents a computational strategy for HEA design, specifically AlCoCrFeNi HEA, that integrates Monte Carlo sampling, molecular dynamics (MD) simulations, and machine learning (ML). A small number of algorithmically chosen molecular simulations in LAMMPS are carried out to create a Gaussian process-based ML model that captures the composition-based variations in Young's modulus and density of the alloys. The prediction-making capabilities of developed ML models are then used to explore the optimal elemental concentrations by utilizing multi-objective genetic algorithm optimization for designing low-density AlCoCrFeNi HEA with enhanced Young's modulus.

High-entropy alloys (HEAs) are a class of multi-principal element alloys comprised of five or more principal alloying elements, each typically present in atomic percentages ranging from 5 to 35% [1, 2]. This unique feature of HEAs in terms of wide compositional space (multiple principal elements) promotes mutual solubility and enhanced configurational entropy. The mutual solubility and high configurational entropy further ensure the formation of solid solutions with the presence of uniform phases [3]. This structural simplicity is responsible for the exceptional properties of HEAs, including robust cryogenic toughness, high strength,

K. K. Gupta et al., *Machine Learning in Nanoscale Materials Design*, Materials Horizons: From Nature to Nanomaterials, https://doi.org/10.1007/978-981-95-2660-4_8

thermal stability at elevated temperatures, and excellent resistance to corrosion and wear [4–7]. Such remarkable mechanical characteristics of HEAs have piqued the interest of the materials science community in harnessing their exceptional mechanical behaviour in multi-faceted applications.

In practice, alloy design can take various approaches, such as modifying element compositions or adjusting the microstructures of existing alloys to achieve specific properties. The process involves fundamental deformation mechanisms, trial-and-error prototyping, high-throughput experiments, computational techniques, and theoretical calculations. However, when exploring the vast compositional space of HEAs to discover new materials, the traditional trial-and-error method used in alloy design can be costly and time-intensive. The application of computational tools can significantly accelerate the alloy design process by predicting desired properties and revealing the underlying mechanisms [8–10]. Therefore, computationally efficient strategies are crucial for quickly identifying alloys with optimal characteristics. Among the different computational approaches, molecular dynamics (MD) is particularly effective and widely used to gain a comprehensive understanding of the microstructural behaviour and mechanical properties of HEAs.

The long-standing challenges in alloy design, such as accurately capturing the composition-driven functional behaviour of HEAs, can be helpful in application-specific programming of the mechanical performance. Additionally, large-scale fabrication and experimentation pose significant challenges. Given these problems associated with the alloy design, it is crucial to developing a computational framework that eliminates the need for extensive experimentation and helps identify the optimal HEA design space with sufficient accuracy. To address these issues, this chapter introduces a computational strategy that effectively integrates random sampling, molecular dynamics (MD) simulations, and machine learning (ML) [11, 12]. This chapter chronologically lays out the step-by-step approach for executing ML-driven molecular dynamics (MD) simulations for guided optimal design of lightweight AlCoCrFeNi HEA for enhanced Young's modulus (E).

8.2 MD Simulation-Driven Dataset

The process of developing a data-driven pipeline for HEA design starts with the generation of initial samples (which must be minimal and sufficient for model construction). For the generation of initial sample space, the Monte Carlo sampling (MCS) strategy is utilized, by randomly generating the concentrations of Al, Fe, Ni, Co, and Cr, within 5% to 35%, with constraints of summation (compositional concentrations should be summed up to 100). MCS is enforced by using the following formulation [13–15]:

$$X_i = X_{\min} + r_{\mathrm{i}} \cdot (X_{\max} - X_{\min}) \tag{8.1}$$

```
import numpy as np
import pandas as pd
# Set random seed for reproducibility
np.random.seed(42)
# Define element names
elements = ['Al', 'Ni', 'Cr', 'Fe', 'Co']
# Define bounds
lower_bound = 5
upper_bound = 35
num_samples = 80  # Number of random samples
# Generate compositions with summation = 100
samples = []
for _ in range(num_samples):
    while True:
        rand_vals = np.random.uniform(lower_bound, upper_bound, len(elements))
        scale_factor = 100 / np.sum(rand_vals)  # Scale to sum to 100
        scaled_vals = rand_vals * scale_factor  # Adjust values
        # Ensure all values remain within [5, 35]
        if np.all(scaled_vals >= lower_bound) and np.all(scaled_vals <= upper_bound):
            samples.append(scaled_vals)
            break
df_random = pd.DataFrame(samples, columns=elements) # Convert to DataFrame
df_random.to_csv('random_samples.csv', index=False) # Save to CSV
```

Fig. 8.1 Python snippet for performing Monte Carlo sampling (MCS). By utilizing MCS, 80 sets of concentration space of AlCoCrFeNi HEA are designed with the constraints for which individual elemental concentration varies from 5 to 35%, and the summation of all the elemental concentration results in 100

where X_{max} and X_{min} refer to the maximum and minimum values for the input features; in our case all the input features (% concentrations of Al, Fe, Co, Cr, and Ni) vary in between 5 and 35%. r_i denotes the random number iteratively generated in between 0 and 1. With the utilization of the MCS scheme, Python code was used for 80 random samples which is given in Fig. 8.1.

The designed alloy samples were computationally modelled (refer to Fig. 8.2) and characterized (uniaxial tensile deformation) within the LAMMPS environment. As an outcome of an individual MD simulation, the density (post-energy minimization) and Young's modulus of different HEA configurations (varied elemental concentration) are recorded.

Such MCS-driven MD simulations resulted in the generation of the dataset required for developing the ML models. The preliminary understanding of the dataset, in terms of inter-elemental correlations and their statistical influence on the desired responses, is obtained through Pearson's correlation analysis (a Python snippet for performing the same is provided in Fig. 8.3). The correlation matrix observed from the dataset is illustrated in Fig. 8.4.

Pearson's correlation revealed that variation in Al concentrations has a strong negative correlation with the target responses (E and ρ), while variation in Ni concentrations depicts positive correlations. The variations in Cr concentrations moderately affect Young's modulus (E) of the AlCoCrFeNi HEA, but have very little effect on its density. All these observations are derived from the dataset, which is obtained by performing MCS-driven MD simulations.

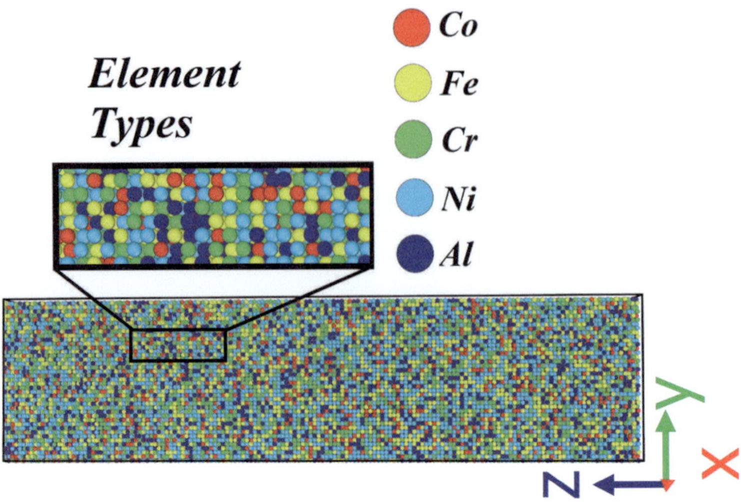

Fig. 8.2 One of the HEA samples modelled in the LAMMPS environment

```python
import pandas as pd
import numpy as np
import matplotlib.pyplot as plt
import seaborn as sns

# Load the dataset from the Excel file
file_path = "/content/Alloy_data.xlsx"
# Read the first sheet of the Excel file
df = pd.read_excel(file_path, sheet_name=0)

# Pearson's Correlation Analysis
plt.figure(figsize=(12, 8), dpi=300)
corr_matrix = df.corr()
sns.heatmap(corr_matrix, annot=True, cmap='jet', fmt='.2f',
            linewidths=0.5,
            cbar_kws={'shrink': 0.8},
            vmax=1.0, vmin=-1.0)
plt.title("Pearson's Correlation Heatmap", fontsize=18, fontweight='bold')
plt.xticks(fontsize=12, rotation=45, ha='right')
plt.yticks(fontsize=12)
plt.savefig("Pearson.png", dpi=300, bbox_inches='tight')
plt.show()
```

Fig. 8.3 Python code snippet for Pearson's correlation plot

8.3 Machine Learning Pipeline

In the next stage, the generated dataset is used to develop the Gaussian process regression (GPR) models [16], capable of making high-fidelity predictions of the responses (E and ρ), for the unknown configurations of AlCoCrFeNi HEA. The GPR

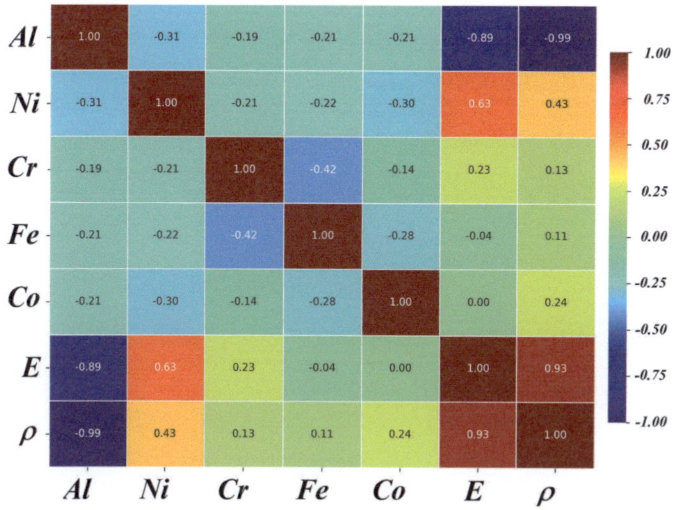

Fig. 8.4 Pearson's correlation plot

models are trained by deploying the ML model training pipeline, offered by the scikit-learn library of Python. The ML pipeline consists of a StandardScaler package for *z*-score normalization of the input features (% concentration of Al, Co, Cr, Fe, and Ni elements), followed by hyperparameter-tuned GPR models by incorporating custom kernel functions (modified radial basis function). The custom function (train_gpr) designed for developing hyperparameter-tuned GPR models is shown in Fig. 8.5. The best GPR models identified by enforcing grid-search cross-validation (fivefold) are used to develop the GPR models for both responses. The best hyperparameter used for model training and the inter-fold model assessment (R^2 value for each fold) are observed by using the Python snippet shown in Fig. 8.6. It is evident from the inter-fold cross-validation scores that obtained R^2 values are consistent across all the folds, regardless of the responses, which highlights the robustness of the prediction-making capability of the developed GPR models.

Once the predictive efficiency of the developed ML models is established, the models are again tested for their generalization capability, by making the predictions for the out-of-the-fold test data. The prediction for the test dataset is compared with the available actual responses (E and ρ), and the performance measures such as R^2 and mean-squared error (MSE) values are calculated (refer to Python code snippet given in Fig. 8.7).

Sample-wise train and test validation of the developed models is illustrated in terms of scatter plots, and the probability distribution of relative error. The distribution of sample points along the linear line of the model demonstrates how close the predictions of the models are with respect to the actual solutions (refer to Fig. 8.8a for E, and Fig. 8.8c for ρ). Similar observations are revealed by the probability distribution function of relative error, corresponding to E (refer to Fig. 8.8b), and ρ

```
from sklearn.pipeline import Pipeline
from sklearn.preprocessing import StandardScaler
from sklearn.metrics import mean_squared_error, r2_score
from sklearn.gaussian_process import GaussianProcessRegressor
from sklearn.gaussian_process.kernels import RBF, ConstantKernel as C
from sklearn.model_selection import train_test_split, GridSearchCV, cross_val_score
# Define features (X) and targets (Y_E for E, Y_ρ for ρ)
X = df[['Al', 'Ni', 'Cr', 'Fe', 'Co']]
Y_E = df['E']
Y_rho = df['ρ']
# Split data into training and testing sets (80% train, 20% test)
X_train, X_test, Y_E_train, Y_E_test = train_test_split(X,
                                              Y_E, test_size=0.1, random_state=42)
X_train, X_test, Y_rho_train, Y_rho_test = train_test_split(X,
                                              Y_rho, test_size=0.1, random_state=42)
# Define a function to train and tune a GPR model
def train_gpr(X_train, Y_train):
    # Define a pipeline with feature scaling and GPR
    pipeline = Pipeline([
        ('scaler', StandardScaler()),
        ('gpr', GaussianProcessRegressor())
    ])
    # Define the hyperparameter grid
    param_grid = {
        'gpr__kernel': [
            C(1.0) * RBF(length_scale) for length_scale in [0.1, 1.0, 10.0]
        ],
        'gpr__alpha': [1e-10, 1e-5, 1e-2]
    }
    # Perform grid search with cross-validation
    grid_search = GridSearchCV(pipeline, param_grid, cv=5,
                        scoring='neg_mean_squared_error', n_jobs=-1)
    grid_search.fit(X_train, Y_train)
    return grid_search.best_estimator_, grid_search.best_params_

# Train and tune the GPR models
best_gpr_E, best_params_E = train_gpr(X_train, Y_E_train)
best_gpr_rho, best_params_rho = train_gpr(X_train, Y_rho_train)
```

Fig. 8.5 Python code snippet for developing hyperparameter-tuned Gaussian process regression models

```
# Display best hyperparameters for both models
best_params_E, best_params_rho

# Cross-validation scores
cv_scores_GPR_E = cross_val_score(best_gpr_E, X_train,
                                  Y_E_train.values.ravel(), cv=5, scoring='r2')
cv_scores_GPR_rho = cross_val_score(best_gpr_rho, X_train,
                                  Y_rho_train.values.ravel(), cv=5, scoring='r2')

print("Best GPR parameters for E:", best_params_E)
print("Best GpR parameters for rho:", best_params_rho)
print("GPR_E 5-fold CV R2 scores:", cv_scores_GPR_E)
print("GPR_rho 5-fold CV R2 scores:", cv_scores_GPR_rho)
```

```
Best GPR parameters for E: {'gpr__alpha': 0.01, 'gpr__kernel': 1**2 * RBF(length_scale=0.1)}
Best GpR parameters for rho: {'gpr__alpha': 1e-05, 'gpr__kernel': 1**2 * RBF(length_scale=1)}
GPR_E 5-fold CV R2 scores: [0.97621715 0.9672563  0.9903158  0.99525446 0.99518011]
GPR_rho 5-fold CV R2 scores: [0.9999072  0.99987824 0.99980953 0.99955492 0.99978152]
```

Fig. 8.6 Python code snippet for obtaining the best hyperparameters used while developing the ML models and the fold-wise R^2 values (*a.k.a.* cv-scores) obtained during hyperparameter tuning

```
# Predictions
y_train_pred_E = best_gpr_E.predict(X_train)
y_test_pred_E = best_gpr_E.predict(X_test)

y_train_pred_rho = best_gpr_rho.predict(X_train)
y_test_pred_rho = best_gpr_rho.predict(X_test)

# Evaluation
r2_train_E = r2_score(Y_E_train, y_train_pred_E)
r2_test_E = r2_score(Y_E_test, y_test_pred_E)
mse_train_E = mean_squared_error(Y_E_train, y_train_pred_E)
mse_test_E = mean_squared_error(Y_E_test, y_test_pred_E)

r2_train_rho = r2_score(Y_rho_train, y_train_pred_rho)
r2_test_rho = r2_score(Y_rho_test, y_test_pred_rho)
mse_train_rho = mean_squared_error(Y_rho_train, y_train_pred_rho)
mse_test_rho = mean_squared_error(Y_rho_test, y_test_pred_rho)

print("GPR_E R-squared (Train, Test):", r2_train_E, r2_test_E)
print("GPR_E MSE (Train, Test):", mse_train_E, mse_test_E)
print("GPR_rho R-squared Train, Test):", r2_train_rho, r2_test_rho)
print("GPR_rho MSE (Train, Test):", mse_train_rho, mse_test_rho)
```

```
GPR_E R-squared (Train, Test): 0.999994233421706 0.9911295406142815
GPR_E MSE (Train, Test): 0.0031321832789540064 2.0843603789123133
GPR_rho R-squared Train, Test): 0.9999783015325783 0.9999496696642843
GPR_rho MSE (Train, Test): 6.606177535869513e-06 7.620922260439951e-06
```

Fig. 8.7 Python code snippet to obtain the performance metrics (R^2 and mean-squared error (MSE)) for training and test dataset, for both the responses (E and ρ)

(refer to Fig. 8.8d). The error analysis reveals that the relative error for both responses is close to zero, regardless of the training and test dataset.

In this manner, the developed GPR models are rigorously tested for prediction accuracy. The rigorous validation of the ML models established that the developed models can be deployed for large-scale predictions. Hence, the exceptional generalization capability of GPR models is further utilized for predicting E and ρ values of unknown configurations of AlCoCrFeNi HEA.

8.4 Deployment of ML Models for Functional Design of HEA

Next, the predictive framework developed in the preceding section is deployed for making large-scale predictions. The main aim of making such predictions is to screen for possible AlCoCrFeNi HEA configurations with enhanced Young's modulus with relatively minimal density. As illustrated in Fig. 8.4, Pearson's analysis between E and ρ suggests a positive correlation between the two. It means with the enhancement in Young's modulus, the density of the alloy also increases. The functional optimality in the design of HEAs needs to meet the trade-off between the maximization of E,

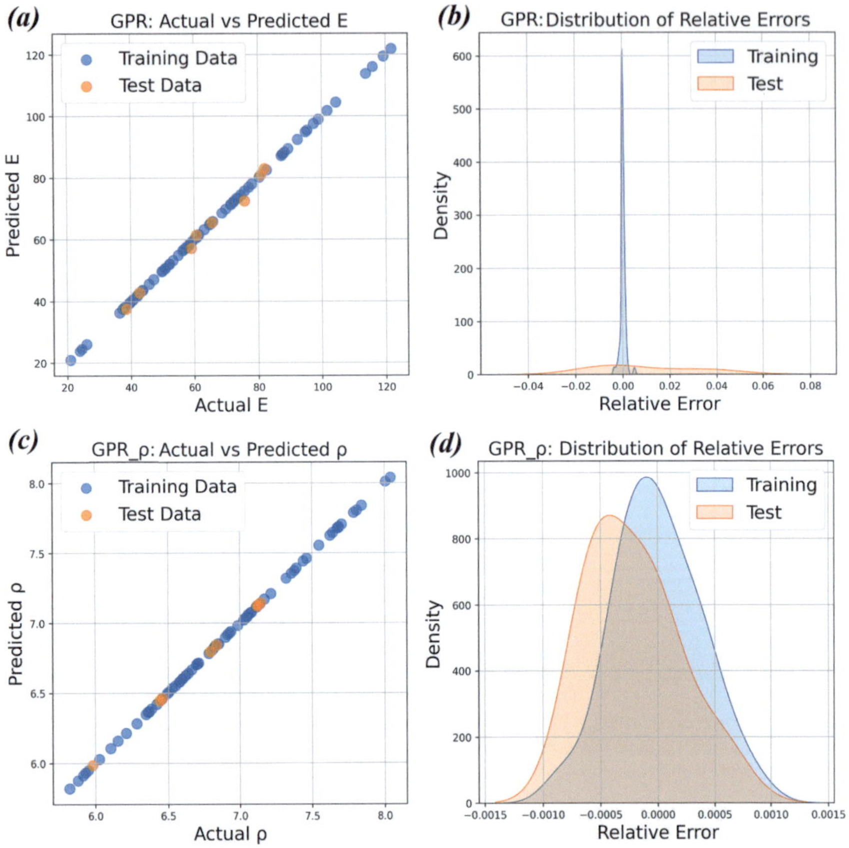

Fig. 8.8 Validation of the developed GPR models: **a** scatter plots for comparing predicted Young's modulus with actual Young's modulus, **b** probability density function (pdf) plots for demonstrating the sample-wise prediction error corresponding to Young's modulus for training and test samples, **c** scatter plots for comparing predicted density with actual density, and **d** probability density function (pdf) plots for demonstrating the sample-wise prediction error corresponding to density for training and test samples

and the minimization of ρ. To obtain this, the developed GPR models are used to explore the Pareto optimal solutions driven by a multi-objective genetic algorithm (MOGA). In this regard, at first, the initial unknown population (with 1000 samples) is generated by using the random sampling method, by enforcing the same constraints which were adopted while generating the base sample space (refer to Fig. 8.1). The 1000 unknown samples were stored as a *dataframe* named df_unknown, from which the samples are taken as a population for enforcing meta-heuristically exploring the Pareto solutions. Within the Python environment, the multi-objective genetic algorithm (MOGA) framework is powered by distributed evolutionary algorithms in Python (DEAP package). A detailed, step-by-step explanation of how the DEAP package is utilized for GPR-driven MOGA is presented below.

4.1. DEAP setup and initial population: The Python code snippet illustrated in Fig. 8.9 installs DEAP, imports necessary libraries, and utilizes the creator function of DEAP to create the individual composition space of alloy, which is randomly initialized from the population selected (500 samples) from randomly generated samples stored in *df_unknown*.
4.2. A custom evaluate function is defined (refer to Fig. 8.10) which utilizes the developed GPR models for making the predictions of E and ρ for the solutions suggested by the MOGA framework.
4.3. The constraints are defined, in terms of summation constraints, custom mutation, and custom crossover.
    4.3.1 The total sum constraint (sum = 100) is enforced to ensure the compositions are physically realistic (refer to Fig. 8.11).
    4.3.2 Custom mutation adds Gaussian noise while keeping the sum fixed and values within specified bounds (refer to Fig. 8.12).
    4.3.3 Custom crossover blends two parent compositions while maintaining the sum constraint (refer to Fig. 8.13).

```
# Set random seed for reproducibility
np.random.seed(42)
# Define element names
elements = ['Al', 'Ni', 'Cr', 'Fe', 'Co']
# Define bounds
lower_bound = 5
upper_bound = 35
num_samples = 1000  # Number of random samples
# Generate compositions with sum = 100
samples = []
for _ in range(num_samples):
    while True:
        rand_vals = np.random.uniform(lower_bound,
                                      upper_bound, len(elements))
        scale_factor = 100 / np.sum(rand_vals)
        scaled_vals = rand_vals * scale_factor

        # Ensure all values remain within [5, 35]
        if np.all(scaled_vals >= lower_bound) and np.all(scaled_vals <= upper_bound):
            samples.append(scaled_vals)
            break
# Convert to DataFrame
df_unknown = pd.DataFrame(samples, columns=elements)
```

Fig. 8.9 DEAP setup and initial population

```
# Function to evaluate objectives
def evaluate(individual):
    x = np.array(individual).reshape(1, -1)  # Convert to 2D array
    # Predict Young's Modulus (E) -> Maximize
    y1_pred = best_gpr_E.predict(x)[0]
    # Predict Density (ρ) -> Minimize
    y2_pred = best_gpr_rho.predict(x)[0]
    return y1_pred, y2_pred  # Return as a tuple of objectives
```

Fig. 8.10 Defining the custom evaluate function using developed GPR models

```
# Define total sum constraint
total_sum = 100  # <-- Add this at the top
toolbox = base.Toolbox()
toolbox.register("individual", tools.initIterate,
                 creator.Individual,
                 lambda: random.choice(selected_population))
toolbox.register("population", tools.initRepeat, list, toolbox.individual)
```

Fig. 8.11 DEAP's toolbox for initialization of individual solutions from the population

```
# **Custom mutation function**
def constrained_mutate(individual, mu=0, sigma=2, indpb=0.2):
    """ Mutate an individual while keeping sum = 100 and values in range """
    new_ind = np.array(individual) + np.random.normal(mu, sigma,
                                                       len(individual))#Gaussian noise
    new_ind = np.clip(new_ind, lower_bound, upper_bound)  # Keep values within bounds
    # Scale to ensure sum = 100
    scaling_factor = total_sum / np.sum(new_ind)
    new_ind = new_ind * scaling_factor
    return creator.Individual(new_ind.tolist()),  # Ensure mutation returns a tuple
toolbox.register("mutate", constrained_mutate)
```

Fig. 8.12 Defining the custom mutation function for ensuring the total summation of solutions to 100, and adding the Gaussian noise to the solutions within the specified bounds

```
def constrained_crossover(ind1, ind2):
    child1 = np.array(ind1)
    child2 = np.array(ind2)
    # Perform a simple blend crossover
    alpha = 0.5
    child1 = alpha * child1 + (1 - alpha) * child2
    child2 = alpha * child2 + (1 - alpha) * child1
    # Scale to ensure sum = 100
    for child in [child1, child2]:
        scaling_factor = total_sum / np.sum(child)
        child *= scaling_factor  # Scale to sum = 100
    return creator.Individual(child1), creator.Individual(child2)
toolbox.register("mate", constrained_crossover)
toolbox.register("select", tools.selNSGA2)
toolbox.register("evaluate", evaluate)
```

Fig. 8.13 Defining the custom crossover function for maintaining the summation constraints within the solutions

4.4. Non-Dominated Sorting Genetic Algorithm (NSGA2) is used to perform multi-objective optimization, which results in a set of solutions offering the trade-off between maximization of E and minimization of ρ. The optimization runs for 50 generations with a population size of 500 and an offspring size of 200. The Pareto solutions (consisting of non-dominated solutions) are sorted and stored as $y1$_values for E and $y2$_values for ρ (Fig. 8.14).

```
# Run the optimization
algorithms.eaMuPlusLambda(population, toolbox, mu=len(population), lambda_=200,
                          cxpb=0.7, mutpb=0.2, ngen=50,
                          stats=None, halloffame=None, verbose=True)

# Extract Pareto front
front = tools.sortNondominated(population, len(population), first_front_only=True)[0]
y1_values = [ind.fitness.values[0] for ind in front]
y2_values = [ind.fitness.values[1] for ind in front]
```

Fig. 8.14 Python code snippet for executing multi-objective optimization powered by GPR-driven NSGA2

4.5. Plot the Pareto Solutions: The Pareto solutions obtained from multi-objective optimization are plotted for the proper visualization trade-off solutions between maximization of E and minimization of ρ (refer to the Pareto front illustrated in Fig. 8.15).

The Pareto solutions obtained from the NSGA2 algorithm are illustrated in Fig. 8.15; the solutions encapsulated in a red elliptical envelope correspond to trade-off solutions which results in sufficient maximization in E and with minimized ρ. It is to be noted that the target responses (E and ρ) present conflicting objective functions, where one cannot be enhanced by enhancing another or vice versa, due to which obtaining trade-off solutions for such a scenario are the best possible way forward to design the lightweight HEA with enhanced Young's modulus. The set of optimal input features (individual concentration of Al, Co, Cr, Fe, and Ni elements) to design HEA with high Young's modulus to density ratio is compared with 2 different solutions available in the base dataset: first, solutions resulting in simultaneous maximization of E and ρ, and, second, solutions resulting in simultaneous

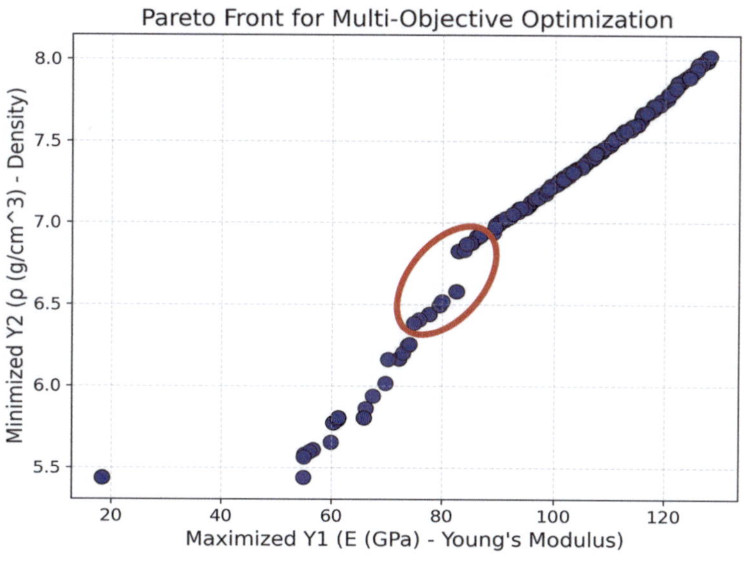

Fig. 8.15 Pareto solutions

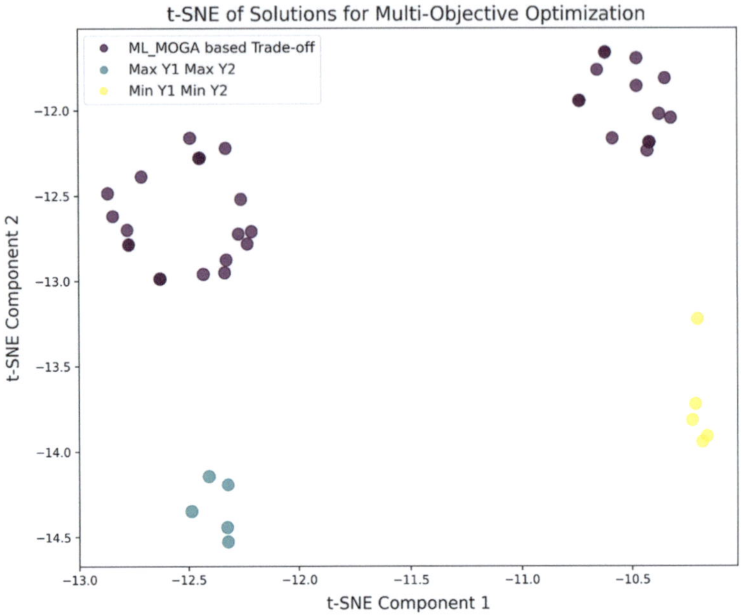

Fig. 8.16 t-SNE of solutions for multi-objective optimization

minimization of E and ρ. The optimal solutions and the base solutions are plotted by using t-distributed Stochastic Neighbour Embedding (t-sne) plots to reveal the distance of optimal solutions from the solutions offered by the base dataset. The t-sne plot is illustrated in Fig. 8.16.

8.5 Summary

In summary, this chapter presents ML-driven optimal design strategy of *AlCoCrFeNi* HEA, wherein the initial dataset is obtained by performing a minimal number of MD simulations for recording the composition-dependent Young's modulus and density of alloy configurations. The chapter laid out the step-by-step hands-on approach for the functional design of high-entropy alloys by utilizing machine learning guided multi-objective genetic algorithm optimization. The outcome of the proposed HEA design pipeline resulted in the individual optimal concentration constituent elements (Al, Co, Cr, Fe, and Ni) which favours an efficient trade-off between maximized Young's modulus and minimized density of the alloys. The outcomes of Chaps. 5 and 6 resulted in a predictive framework capable of making large-scale high-fidelity predictions, while the ML-based pipeline presented in this chapter showcases the effective use of such prediction pipeline for optimal design of compositional materials.

8.6 How to Use the Source Files Associated with This Chapter?

The source files needed to reproduce the investigation laid out in this chapter are available in the GitHub repository linked at https://github.com/KriteshKG/ML4AtomisticSimulations/tree/main/Chapter_8

The link consists of two different folders, one for conducting the MD simulations and the other folder for ML model formation.

8.6.1 MD Simulation

The folder pertaining to MD simulations (named MD_Simulations) consists of the required files to perform the nanoscale simulation of uniaxial tensile deformation over a designed configuration of high-entropy alloy. The MD_Simulations folder namely consists of the following files:

LAMMPS input script:	*input.in*
Interatomic potential:	*_FeNiCrCoAl.eam.alloy*

The set of LAMMPS input scripts, and IP files, can be used to perform the MD simulation. To run the simulation, the terminal is to be opened in the directory, and the following string of instructive code is to be entered:

lmp -in input.in -pk omp 4 -sf omp

The MD simulations result in a text file (_stressvstrain2.txt) which records the increment in strain and the corresponding stress in Z (along the loading direction) which can be used to deduce the stress–strain profile of the graphene configurations, in terms of Young's modulus and density of the compositionally varied HEAs.

8.6.2 Machine Learning

Another folder in the provided GitHub repository, named Machine_Learning, consists of two files: first, the Excel file (Alloy_data.xlsx) which is the dataset for the discussed regression problem, and, second, the Python notebook (Chapter_8.ipynb) for executing the ML model-based investigation reported in this chapter. This notebook can be opened within the Google COLAB environment or by using Jupyter Notebooks on the local machine. Other necessary Excel files used in the Python notebook are also provided in this folder.

References

1. Yeh JW, Chen SK, Lin SJ, Gan JY, Chin TS, Shun TT, Tsau CH, Chang SY (2004) Nanostructured high-entropy alloys with multiple principal elements: novel alloy design concepts and outcomes. Adv Eng Mater 6(5):299–303
2. Tokarewicz M, Grądzka-Dahlke M (2021) Review of recent research on AlCoCrFeNi high-entropy alloy. Metals 11(8):1302
3. Yeh JW (2016) Overview of high-entropy alloys. In: High-entropy alloys: fundamentals and applications, pp 1–19
4. Gludovatz B, Hohenwarter A, Catoor D, Chang EH, George EP, Ritchie RO (2014) A fracture-resistant high-entropy alloy for cryogenic applications. Science 345(6201):1153–1158
5. Chou YL, Wang YC, Yeh JW, Shih HC (2010) Pitting corrosion of the high-entropy alloy Co1.5CrFeNi1.5Ti0.5Mo0.1 in chloride-containing sulphate solutions. Corros Sci 52(10):3481–3491
6. Wu YD, Cai YH, Wang T, Si JJ, Zhu J, Wang YD, Hui XD (2014) A refractory Hf25Nb25Ti25Zr25 high-entropy alloy with excellent structural stability and tensile properties. Mater Lett 130:277–280
7. Yu Y, Wang J, Li J, Kou H, Duan H, Li J, Liu W (2015) Tribological behavior of AlCoCrCuFeNi and AlCoCrFeNiTi0.5 high entropy alloys under hydrogen peroxide solution against different counterparts. Tribol Int 92:203–210
8. Barman S, Kumar Gupta K, Dey S (2024) Exploring nano-scale scratching induced tribological behavior of graphene engineered AlCoCrFeNi high-entropy alloy. J Appl Mech 91(10)
9. Barman S, Gupta KK, Dey S (2024) Probing atomistic deformation behavior of graphene-coated Al0. 3CoCrFeNi high-entropy alloy under nanoindentation. J Micromech Molec Phys 9(01):1–14
10. Barman S, Gupta KK, Dey S (2024) Enhancing mechanical performance of Al0. 3CoCrFeNi HEA films through graphene coating: insights from nanoindentation and dislocation mechanism analysis. Modell Simul Mater Sci Eng 32(3):035012
11. Gupta KK, Barman S, Dey S, Naskar S, Mukhopadhyay T (2024) On exploiting nonparametric kernel-based probabilistic machine learning over the large compositional space of high entropy alloys for optimal nanoscale ballistics. Sci Rep 14(1):16795
12. Gupta KK, Barman S, Dey S, Mukhopadhyay T (2024) Explainable machine learning assisted molecular-level insights for enhanced specific stiffness exploiting the large compositional space of AlCoCrFeNi high entropy alloys. Mach Learn: Sci Technol 5(2):025082
13. Gautam, Gupta KK, Bhowmik D, Dey S (2023) Probing the stochastic unconfined compressive strength of lime–RHA mix treated clayey soil. J Mater Civ Eng 35(3):04022469
14. Roy A, Gupta KK, Dey S (2022) Probabilistic investigation of temperature-dependent vibrational behavior of hetero-nanotubes. Appl Nanosci 12(7):2077–2089
15. Mukhopadhyay T, Naskar S, Gupta KK, Kumar R, Dey S, Adhikari S (2021) Probing the stochastic dynamics of coronaviruses: machine learning assisted deep computational insights with exploitable dimensions. Adv Theory Simul 4(7):2000291
16. Mahata A, Mukhopadhyay T, Chakraborty S, Asle Zaeem M (2024) Atomistic simulation assisted error-inclusive Bayesian machine learning for probabilistically unraveling the mechanical properties of solidified metals. NPJ Comput Mater 10(1):22

Chapter 9
High-Entropy Alloy: Optimal Nanoscale Ballistic Performance

9.1 Overview

In Chap. 8, the AlCoCrFeNi HEA configurations were designed to meet the Pareto optimality criteria for enhancing Young's modulus while minimizing the density. The Pareto optimality offers a best-suited trade-off between the objective functions so that the decision variables can be properly tuned within the envelope of parametric variations. In this chapter, the ML-driven optimization framework is integrated with the high-velocity impact characterization of AlCoCrFeNi HEA. The best ML model suggested by Python's LazyPredict is deployed to perform the Bayesian optimization to screen for the best likely solutions (optimal composition of AlCoCrFeNi HEA) that can result in enhanced kinetic energy absorption against high-velocity impact.

Due to their extensive compositional range and exceptional attributes, including high hardness, ultimate tensile strength, ductility, and enhanced resistance to corrosion and wear, HEAs offer significant potential for a wide array of applications. As a result, HEAs are increasingly utilized in aerospace, marine, high-temperature, and cryogenic industries to leverage these advanced properties. Recent efforts to explore materials for high-velocity impact have revealed that HEAs exhibit promising potential for designing and developing materials capable of withstanding such impacts [1, 2]. In such applications, materials are exposed to high-velocity impacts that induce significant strain deformation, with strain rates exceeding approximately 10^2 s^{-1}.

The published literature indicates that single-phase HEA systems exhibit exceptional strain rate sensitivity and work hardening capabilities, attributed to their low stacking fault energy (SFE). On the other hand, dual-phase HEA systems demonstrate enhanced strength due to a refined microstructure, which includes a large number of interphase boundaries. Recent studies have also highlighted the promising properties of AlCoCrFeNi HEA in withstanding dynamic impact and high strain rate deformations. Therefore, it is crucial to assessing the ballistic behaviour of such materials, considering the large-scale variations in their compositional space [3]. However,

K. K. Gupta et al., *Machine Learning in Nanoscale Materials Design*, Materials Horizons: From Nature to Nanomaterials, https://doi.org/10.1007/978-981-95-2660-4_9

conducting such assessments experimentally is challenging due to time, cost, and precision constraints.

In contrast, molecular dynamics (MD) simulations have proven effective in providing insights into material behaviour at the atomic level with sufficient accuracy. To overcome the limitations of experimental methods, several research groups have employed MD simulations to investigate the nanoscale ballistic performance of potential barrier materials [4–6]. Despite this, characterizing the ballistic performance of HEAs across a broad compositional space remains a significant challenge.

To address this challenge, the present chapter proposes the integration of machine learning (ML) approaches with traditional MD simulations, as demonstrated in previous studies, to capture continuous variations in parametric settings. By combining MD simulations with a computationally efficient ML model, the ballistic response (kinetic energy dissipation, ΔKE) of HEA samples with varied compositions can be mapped according to changes in the alloying composition of individual elements. Furthermore, this combined MD-ML framework is integrated with Bayesian optimization for the ML-guided design of AlCoCrFeNi HEA to withstand high-velocity impact, enabling the optimal tailoring of its properties.

9.2 MD Simulation-Driven Dataset

At first, the input features, such as the atomic fractions of the constituent elements (Al, Co, Cr, Fe, and Ni) in the HEA, are randomly distributed within the defined parametric range (as shown in Table 9.1) using random sampling (refer to Eq. 9.1 [7–10]). This process creates the sample space for the machine learning model:

$$X_i = X_{min} + r_i \cdot (X_{max} - X_{min}) \tag{9.1}$$

where X_{max} and X_{min} refer to the maximum and minimum values for the input features; in our case, all the input features (% concentrations of Al, Fe, Co, Cr, and Ni) vary in between 5 and 35%. r_i denotes the random number iteratively generated in between 0 and 1.

Table 9.1 Parametric range of variation of input features

Input features	Variation range (%)
Fe	15–25
Ni	15–25
Co	15–25
Cr	15–25
Al	0–40

With the utilization of the MCS scheme, 80 numbers of random samples are generated. Python code snippet for the same is illustrated in Fig. 8.1 of Chap. 8. Next, MD simulations of high-velocity impact are carried out on various configurations of AlCoCrFeNi HEA, with the randomly distributed input sample space applied. To simulate the projectile impact (with an impact velocity of 7.5 Å/ps) in the LAMMPS environment, the AlCoCrFeNi HEA with the random compositional variations is modelled as a cylindrical target plate (radius = 215 Å, thickness = 60.52 Å). A 15 Å-wide section at the target plate's periphery is fixed by applying zero force and zero velocity. A rigid cylindrical gold projectile (radius = 28 Å, height = 78 Å) is used for the atomistic impact simulations. The interactions between the elements of the AlCoCrFeNi HEA are modelled using the *EAM/alloy* force field, while the gold nano-projectile is modelled with the *EAM* force field. MD simulations of high-velocity impact are carried out with an iteration time step of 0.1 fs. Before the high-velocity impact simulations begin, the potential energy of the system is minimized for 2.5 ps. The NPT ensemble is applied for 10 ps to equilibrate the system's pressure and temperature to 0 bar and 300 K, respectively. Finally, the projectile is simulated impacting the target plate within the NVE ensemble.

The instantaneous variation in the kinetic energy of the projectile is analysed to determine the kinetic energy dissipated (ΔKE) by the target plate, which can be calculated as follows:

$$\Delta\mathrm{KE} = \mathrm{KE}_{\mathrm{i}} - \mathrm{KE}_{f} \tag{9.2}$$

where KE_{i} stands for the initial kinetic energy of the projectile and KE_{f} denotes the post-impact kinetic energy of the projectile. The dataset generated by such random sampling-based MD simulation is utilized to train (with 90%) and validate (with separate 10% samples) the ML model. Pearson's correlation of the observed MD simulation dataset is illustrated in Fig. 9.1 (Python code snippet provided in Fig. 8.3 of Chap. 8).

It is evident from Pearson's correlation analysis presented in Fig. 9.1 that the presence of high concentrations of Cr, Fe, and Co elements enhances the kinetic energy absorption, highlighting the capacity to withstand high-velocity impact. In contrast, an increase in Al concentration demonstrates negative influence, and the Ni concentration does not influence the ballistic response (as showing 0 correlation).

9.3 Machine Learning Pipeline

In the next stage, the generated dataset is used to develop a best-suited ML model, suggested by *LazyPredict* in Python (snippet of the code illustrated in Fig. 5.5 of Chap. 5). The superficial comparison of ML model performance is provided in Fig. 9.2, which highlights the superior predictive performance offered by the Extreme Gradient Boosting (XGBoost) ML model (with R^2 as 0.99, and RMSE as 55.05). With the preliminary understanding of the ML model that can be used to develop

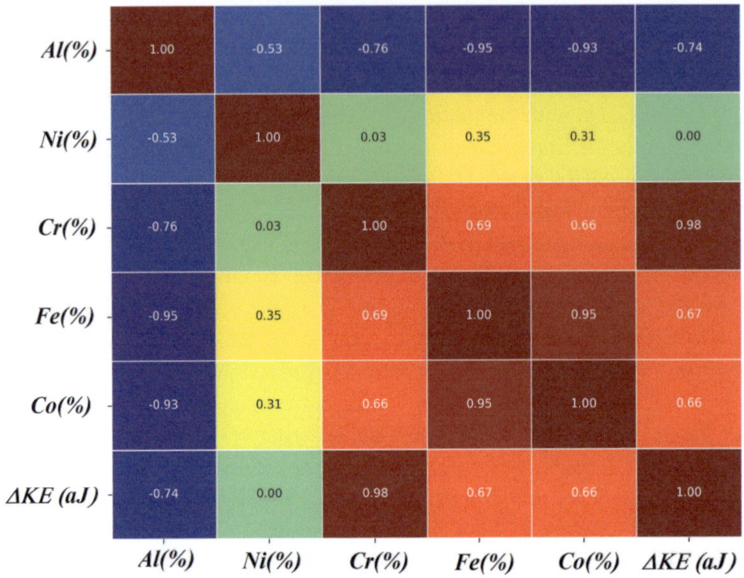

Fig. 9.1 Pearson's correlation plot for illustrating the statistical significance of an individual constituent element on the ballistic performance (kinetic energy absorption ΔKE (*aJ*)) of AlCoCrFeNi HEA

the computationally efficient predictive model, in the next stage, the XGBoost-based ML pipeline for property prediction is prepared. The ML pipeline chronologically implemented *z*-score normalization (through the scikitlearn library's StandardScaler package) to the input features followed by the development of XGBRegressor by using a training dataset. The ML model development utilized hyperparameter tuning by incorporating the GridSearch *k*-fold cross-validation ($k = 5$) scheme.

The hyperparameters of the XGBoost ML model, such as number of estimators, learning rate, maximum depth, subsample (fraction of training data to be used), column samples per tree (fraction of features used during tree building), and regularization term (gamma), are explored for developing the best performing XGBoost regressor, by using an optimal hyperparameter combination. Python code snippet of the utilized ML pipeline is illustrated in Fig. 9.3.

The best hyperparameters used to train the XGBoost ML model and fold-wise (fivefold) R^2 values are illustrated in Fig. 9.4.

It is evident from Fig. 9.4 that the developed ML models demonstrated consistently higher R^2 values across the 5 folds. The ML validation can be further observed through scatter plots and probability density function (pdf) plots (relative error) for training and test data (illustrated in Fig. 9.5).

The ML validation illustrated in Fig. 9.5 demonstrates that the developed model has exceptional generalization capability and is capable of predicting ballistic response with minimal error.

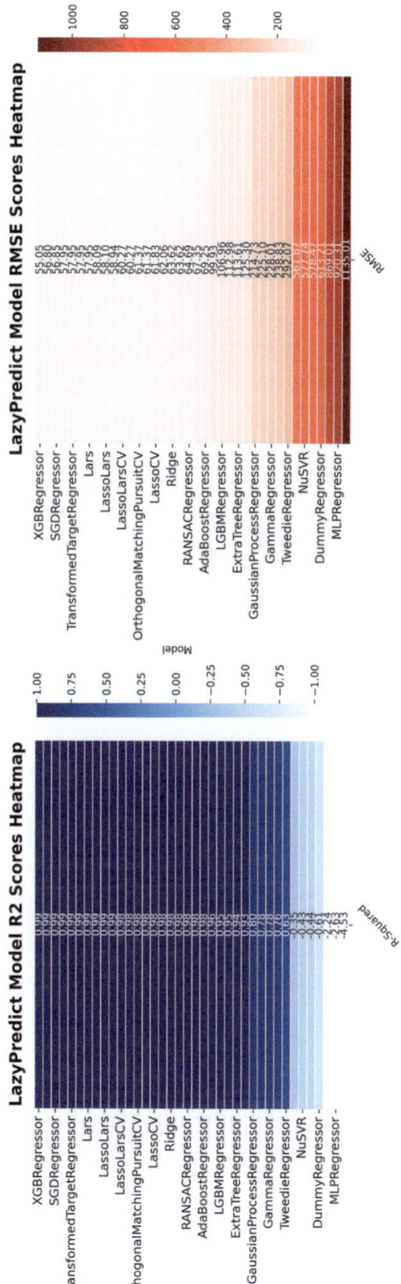

Fig. 9.2 Comparison of ML model performance offered by *LazyPredict*

```python
from xgboost import XGBRegressor
from sklearn.pipeline import Pipeline
from sklearn.preprocessing import StandardScaler
from sklearn.model_selection import GridSearchCV
def train_xgboost(X_train, Y_train): # A function to train XGBoost regressor model
    pipeline = Pipeline([
        ('scaler', StandardScaler()),
        ('xgboost', XGBRegressor())
    ])
    # Define the hyperparameter grid for XGBoost
    param_grid = {
        'xgboost__n_estimators': [100, 200, 300],
        'xgboost__learning_rate': [0.01, 0.05, 0.1],
        'xgboost__max_depth': [3, 5, 7],
        'xgboost__subsample': [0.8, 1.0],
        'xgboost__colsample_bytree': [0.8, 1.0],
        'xgboost__gamma': [0, 0.1, 0.2]
    }
    grid_search = GridSearchCV(pipeline, param_grid, cv=5,
                               scoring='neg_mean_squared_error', n_jobs=-1)
    grid_search.fit(X_train, Y_train)
    return grid_search.best_estimator_, grid_search.best_params_
```

Fig. 9.3 XGBoost regressor pipeline for developing hyperparameter-tuned ML model

```python
# Best Hyperparameter and CV Scores
best_model, best_params = train_xgboost(X_train, y_train)
cv_scores = cross_val_score(best_model, X_train, y_train.values.ravel(), cv=5, scoring='r2')
print("Best hyperparameters:", best_params)
print("5-fold CV R2 scores:", cv_scores)
```

```
Best hyperparameters: {'xgboost__colsample_bytree': 1.0,
                       'xgboost__gamma': 0,
                       'xgboost__learning_rate': 0.05,
                       'xgboost__max_depth': 3,
                       'xgboost__n_estimators': 100,
                       'xgboost__subsample': 0.8}
5-fold CV R2 scores: [0.9831404  0.90418454 0.94420097 0.98386233 0.95515622]
```

Fig. 9.4 Best hyperparameters of XGBoost ML model and R-squared values of ML models across the cross-folds

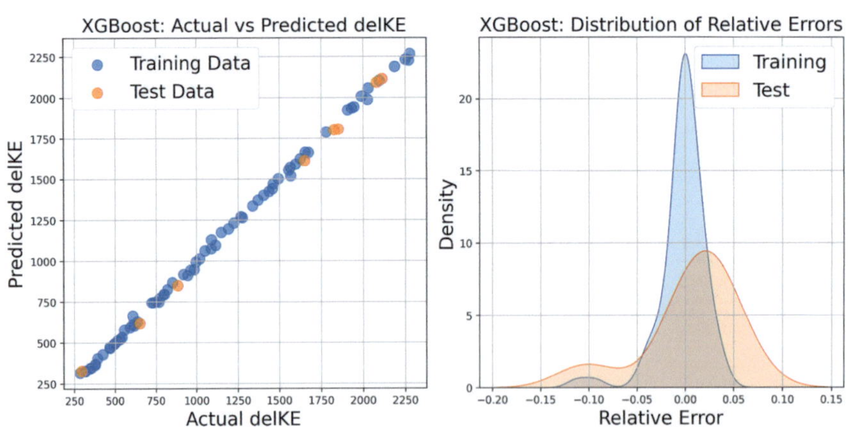

Fig. 9.5 Scatter plots comparing actual kinetic energy absorption (MD simulation-based delKE) and predicted kinetic energy absorption for train and test datasets, and the pdf of relative error for training and test datasets

9.4 Deployment of ML Model for Bayesian Optimization

Next, the developed XGBoost ML model is deployed for performing Bayesian optimization to explore the optimal solution (optimal concentration of constituent elements of AlCoCrFeNi HEA) for enhancing kinetic energy absorption of HEA. In this regard, at first, the virtual alloy samples (1000 samples) are generated by using the random sampling approach, wherein the elemental concentrations are varied within the stipulated parametric range of variation (refer to Table 9.1) and enforced with the summation constraint. The snippet of the Python code for random sampling of the AlCoCrFeNi HEA is provided in Fig. 8.1 of Chap. 8. The generated virtual alloy samples (1000 samples) function as a population of solutions, in which Bayesian exploration is to be performed to screen the best alloy composition for withstanding high-velocity impact. The ML-guided Bayesian optimization of AlCoCrFeNi HEA executes two critical operations: first, bootstrapped training of ML models, capturing the uncertainty in predictions of the bootstrapped ML models, and, second, calculation of expected improvement (EI) score to find the expected optimal configurations of HEA for maximized ballistic performance. The chronologically laid out methodology and observations of Bayesian optimization are as follows.

9.4.1 Bootstrapping

At first, the training and test datasets are combined into a single dataset (*X_train_all*, *y_train_all*), to use the complete dataset as a whole in the training process during bootstrapped ML model formation. Bayesian optimization also considers the unknown population (randomly sampled 1000 HEA configurations), during the process (stored as '*df_unknown*'). Hence, in the next step the compiled input features (*X_train_all*) and unknown population (*df_unknown*) are z-score-normalized using the *StandardScaler* package of *scikitlearn* library (refer to snippet of Python code in Fig. 9.6).

Next, a function named '*bootstrap_estimator*' is defined to implement bootstrapping, which is a technique for estimating the performance of a model by resampling the training data. The ML model is repeatedly trained on bootstrapped samples (randomly sampled with replacement) from the training data. For each bootstrapped sample space, the model makes predictions on test data (refer to a snippet of Python code in Fig. 9.7).

```
#Normalizing df_unknown
X_train_all = np.concatenate((X_train,X_test),axis = 0)
y_train_all = np.concatenate((y_train,y_test),axis=0)

scaler = StandardScaler()
X_train_all_scaled = scaler.fit_transform(X_train_all) ## fit and transform all train data
X_unknown_scaled  = scaler.transform(df_unknown) # transform unknown 1000 samples
```

Fig. 9.6 Preparing the dataset for bootstrapping

```
from scipy.stats import norm
def bootstrap_estimator(model, X_train, y_train, X_test, n_iter=100):
  bootstrap_preds = np.zeros([len(X_test), n_iter])
  index = np.arange(X_train.shape[0])
  np.random.seed(20)  # Fix seed for reproducibility
  for i in range(n_iter):
      index_sampled = np.random.choice(index, size=X_train.shape[0], replace=True)
        # Add small Gaussian noise to prevent identical predictions
        # Use array indexing instead of .iloc
      X_train_sample = X_train[index_sampled, :] + np.random.normal(0, 0.01, X_train.shape)
      y_train_sample = np.array(y_train)[index_sampled].reshape(-1, 1)
      model.fit(X_train_sample, y_train_sample.ravel())
      bootstrap_preds[:, i] = model.predict(X_test)
  return bootstrap_preds.mean(1), bootstrap_preds.std(1)
```

Fig. 9.7 Bootstrapping function

Here, iteratively bootstrapped samples are utilized to train 100 (reflected by '*n_iter* = 100') different models, and with each model prediction on test data is made, which enables capturing mean of predictions '*mu_x*' across all iterations, and standard deviation ('*sigma_x*') of predictions across all iterations. In summary, the bootstrapping function results in the uncertainty in the model's prediction.

9.4.2 Expected Improvement (EI) Score Calculation

The outcomes of bootstrapped ML model formation are used to calculate the expected improvement (EI) score by using the following equation:

$$\begin{aligned}
&\mathrm{EI}(x) = (mu(x) - mu_{\max} - x_i) \times \Phi(z) + \sigma(x) \times \phi(z)\\
&\text{where}\\
&mu(x) = \text{mean of predictions } x\text{th iteration}\\
&\sigma(x) = \text{standard deviation of predictions } x\text{th iteration}\\
&mu_{\max} = \text{95th percentile of the observed training data values(y_ train_ all)}\\
&z = \left(\frac{mu(x) - mu_{\max}}{\sigma(x)}\right)\\
&\Phi(z) = \text{CDF(Cumulative Distribution Function) of improvement}\\
&\quad \text{in mean of the predictions}\\
&\phi(z) = \text{PDF(Probability Density Function) of improvement in}\\
&\quad \text{mean of the predictions.}
\end{aligned} \tag{9.3}$$

The Python code snippet for the function defined to calculate the EI scores is illustrated in Fig. 9.8.

It is to be noted from Fig. 9.8 that the defined function *Expected_Improvement* calls the function *bootstrap_estimator* for applying the *best_model* (developed XGBoost regressor) to calculate the EI score (refer to Fig. 9.9). The EI score calculation also considers the exploration parameter ($x_i = 0.1$), which controls the balance

```python
# Expected Improvement function
def Expected_Improvement(X_test, X_train_all, y_train_all, model, xi=0.1):
  mu_x, sigma_x = bootstrap_estimator(model, X_train_all, y_train_all, X_test, n_iter=10000)
  epsilon = 1e-6  # Small offset to prevent division by zero
  sigma_x = np.where(sigma_x == 0, epsilon, sigma_x)
  mu_max = np.percentile(y_train_all, 95)  # Use 95th percentile instead of max
  diff = mu_x - mu_max - xi
  z = diff / sigma_x
  ei = np.where(sigma_x > 0, diff * norm.cdf(z) + sigma_x * norm.pdf(z), 0)
  return ei, mu_x, sigma_x
```

Fig. 9.8 Snippet of the Python code defining the function for expected improvement (EI) score calculation

```python
# Run EI calculation
ei, predicted_property, std = Expected_Improvement(X_unknown_scaled,
                                                   X_train_all_scaled,
                                                   y_train_all,
                                                   best_model, xi=0.1)
# Create DataFrame
df_predict = pd.DataFrame({"ei": ei, "predicted_property": predicted_property, "std": std})
# Concatenate df_unknown (1000 unknown samples) and df_predict (ei_scores)
df_concatenated = pd.concat([df_unknown, df_predict], axis=1)
# Sort results (for highest EI score)
df_predict_final_sorted = df_concatenated.sort_values('ei', ascending=False)
df_predict_final_sorted.to_csv('/content/ei_sorted.csv')
# Display top results
print(df_predict_final_sorted.head(10))
```

Fig. 9.9 The snippet of the Python code for calculating and post-processing EI scores

between exploration and exploitation. The higher value of x_i results in an increase in exploration.

Next, the calculated EI scores and the corresponding prediction of kinetic energy absorption (delKE) for unknown 1000 samples, and the standard deviation in the predictions are concatenated and stored in a *dataframe* named '*df_concatenated*' (refer to Fig. 9.9). The concatenated *dataframe* is then sorted on the basis of EI scores (based on descending order) to highlight the compositions favouring the maximization in the ballistic response. The first ten optimal solutions obtained on the basis of the highest EI scores are compared with the maximum delKE solutions (present in the base dataset), and minimum delKE solutions (present in base dataset) by using the t-SNE dimensionality reduction plot (illustrated in Fig. 9.10). The optimal solutions and the base solutions are plotted by using t-distributed Stochastic Neighbour Embedding (t-sne) plots to reveal the distance of optimal solutions from the solutions (maximum delKE) offered by the base dataset.

9.5 Summary

In summary, this chapter presents ML-driven optimal design strategy of AlCoCrFeNi HEA, wherein the initial dataset is obtained by performing a minimal number of MD simulations for recording the composition-dependent ballistic (high-velocity) impact performance of the HEA in terms of kinetic energy dissipation (delKE). The chapter

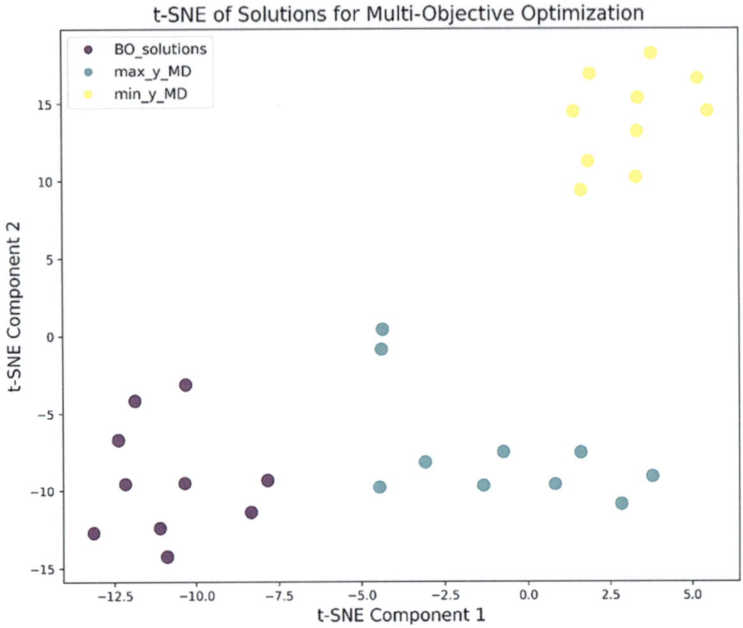

Fig. 9.10 t-SNE of solutions for Multi-Objective Optimization

laid out a step-by-step hands-on approach for the functional design of high-entropy alloys by utilizing machine learning-guided Bayesian optimization. The outcome of the proposed HEA design pipeline resulted in the individual optimal concentration constituent elements (Al, Co, Cr, Fe, and Ni) which favours an enhanced ballistic performance. The proposed ML-driven Bayesian optimization framework presents an active learning approach to materials design, which helps in the cases where an initial dataset is not in sufficient numbers.

9.6 How to Use the Source Files Associated with This Chapter?

The source files needed to reproduce the investigation laid out in this chapter are available in the GitHub repository linked at https://github.com/KriteshKG/ML4AtomisticSimulations/tree/main/Chapter_9

The link consists of two different folders, one for conducting the MD simulations and the other folder for ML model formation.

9.6.1 MD Simulation

The folder pertaining to MD simulations (named MD_Simulations) consists of the required files to perform the nanoscale simulation of uniaxial tensile deformation over a designed configuration of high-entropy alloy. The MD_Simulations folder namely consists of the following files:

LAMMPS input script:	*input.in*
Interatomic potential: for modelling HEA	*FeNiCrCoAl.eam.alloy*
Interatomic potential: for modelling Au projectile	*Au_5T.eam.alloy*
Datafile:	*Fe_29.57Cr_12.71Ni_7.2Co_40.47Al_10.xyz*

The set of LAMMPS input scripts, datafile, and IP files can be used to perform the MD simulation. To run the simulation, the terminal is to be opened in the directory, and the following string of instructive code is to be entered:

lmp -in input.in -pk omp 4 -sf omp

The MD simulations result in a text file (output_impactor.txt) which records the temporal variation in kinetic energy of the impactor, which is used to obtain the pre-impact and post-impact kinetic energy of the projectile, which is then used to calculate kinetic energy dissipation (ΔKE) by utilizing Eq. 9.2.

9.6.2 Machine Learning

Another folder in the provided GitHub repository, named Machine_Learning, consists of two files: first, the Excel file (Impact_Dataset.xlsx) which is the dataset for the discussed regression problem, and, second, the Python notebook (Chapter_9.ipynb) for executing the ML model-based investigation reported in this chapter. This notebook can be opened within the Google COLAB environment or by using Jupyter Notebooks on the local machine. Other necessary Excel files used in the Python notebook are also provided in this folder.

References

1. Choudhuri D, Jannotti PA, Muskeri S, Shukla S, Gangireddy S, Mukherjee S, Schuster BE, Lloyd JT, Mishra RS (2019) Ballistic response of a FCC-B2 eutectic AlCoCrFeNi 2.1 high entropy alloy. J Dyn Behav Mater 5:495–503
2. Choudhuri D, Shukla S, Jannotti PA, Muskeri S, Mukherjee S, Lloyd JT, Mishra RS (2019) Characterization of as-cast microstructural heterogeneities and damage mechanisms in eutectic AlCoCrFeNi2.1 high entropy alloy. Mater Charact 158:109955

3. Gupta KK, Barman S, Dey S, Naskar S, Mukhopadhyay T (2024) On exploiting nonparametric kernel-based probabilistic machine learning over the large compositional space of high entropy alloys for optimal nanoscale ballistics. Sci Rep 14(1):16795
4. Gupta KK, Roy L, Dey S (2022) Ballistic performance of bi-layer graphene: artificial neural network based molecular dynamics simulations. In: Machine learning applied to composite materials. Springer Nature Singapore, Singapore, pp 135–148
5. Gupta KK, Mukhopadhyay T, Roy L, Dey S (2022) High-velocity ballistics of twisted bilayer graphene under stochastic disorder. Adv Nano Res 12(5):529–547
6. Dewapriya MAN, Miller RE (2021) Energy absorption mechanisms of nanoscopic multilayer structures under ballistic impact loading. Comput Mater Sci 195:110504
7. Gautam, Gupta KK, Bhowmik D, Dey S (2023) Probing the stochastic unconfined compressive strength of lime–RHA mix treated clayey soil. J Mater Civ Eng 35(3):04022469
8. Roy A, Gupta KK, Dey S (2022) Probabilistic investigation of temperature-dependent vibrational behavior of hetero-nanotubes. Appl Nanosci 12(7):2077–2089
9. Mukhopadhyay T, Naskar S, Gupta KK, Kumar R, Dey S, Adhikari S (2021) Probing the stochastic dynamics of coronaviruses: machine learning assisted deep computational insights with exploitable dimensions. Adv Theory Simul 4(7):2000291
10. Gupta KK, Barman S, Dey S, Mukhopadhyay T (2024) Explainable machine learning assisted molecular-level insights for enhanced specific stiffness exploiting the large compositional space of AlCoCrFeNi high entropy alloys. Mach Learn: Sci Technol 5(2):025082

Chapter 10
Materials Informatics

10.1 Overview

The previous chapters of this book explored the integration of machine learning frameworks with the quasi-static and dynamic characterization of nanomaterials, such as graphene and high-entropy alloys (specifically from Chaps. 5 *to* 9*). The development of these predictive frameworks enables large-scale characterization of material systems and subsequently leads to virtual screening. Despite the effectiveness of these predictive pipelines, several challenges remain. These include the generation of structured datasets, the scalability of machine learning pipelines, and the need to instil domain-specific awareness in the developed models. To address these challenges, this final chapter of the book presents several practical solutions for developing a materials informatics pipeline aimed at materials design and discovery.*

The Materials Genome Initiative (MGI) proposed by the US Government in 2011 [1] presented the notion of adapting to data-driven high-throughput methods for accelerating the materials discovery, design for specific applications, and deployment. Such an initiative was timely and essential, in light of the slow pace of breakthroughs in materials design that were observed in the past. This happened due to a mix of limited early scientific knowledge, economic resistance to change, and strict industrial policies [2]. The materials genome initiative aims to deploy computational frameworks (simulations backed up with artificial intelligence) to unravel the materials genome (structure–property relationship) which would lead to rapid technological advancements with minimal expense of time and resources [3]. This attracted the interest of the materials research community in constructing integrated computational frameworks capable of utilizing the least amount of exploration while capitalizing on the researched discoveries through computational intelligence [4–8]. Computationally efficient, informatics-driven frameworks exhibit the promising potential to significantly reduce the time and resources required for materials discovery compared to traditional lab methods. Despite their promising potential in accelerating the design and discovery of materials, data-driven approaches often face functional challenges

K. K. Gupta et al., *Machine Learning in Nanoscale Materials Design*, Materials Horizons: From Nature to Nanomaterials,
https://doi.org/10.1007/978-981-95-2660-4_10

that can hinder their implementation. In this context, this chapter explores practical solutions to help mitigate these challenges.

10.2 Generation of Structured Dataset

Access to the dataset is required for the construction of a data-driven materials design pipeline; however, getting such a large dataset through atomistic simulations is typically not practicable. The 'curse of data unavailability' in the materials science domain makes it difficult to utilize ML-driven frameworks. To address such limitations, the materials research community has started sharing the dataset on different open-access platforms. The open-access journals such as Data in Brief (Elsevier), Open Data (Springer), and Data (MDPI) can be of great help in exploring the experimental or simulation-based dataset for desired materials. Despite such options, a significant challenge remains in finding high-throughput, domain-specific datasets for targeted materials systems. The root of this issue lies in the resource-intensive nature of generating such datasets, for instance, performing large-scale molecular dynamics (MD) simulations. To help overcome these limitations, this section highlights several open-source repositories that provide valuable data for a wide range of material classes.

The online data repository such as Materials Project (MP) [9] offers compositional, structural, and functional data on materials such as catalysts, perovskites, battery materials, and metal–organic frameworks. Through its Python-based Application Programming Interface (API), MP enables users to query different classes of materials—an essential feature for developing predictive frameworks. For instance, Ong et al. [10] developed the REpresentational State Transfer (REST) API to facilitate querying with MP's database. Such utilization of an MP data repository resulted in several studies: Dung et al. [11] developed a computational framework for analysing X-ray diffraction patterns, while Liu et al. [12] used MP data to build a machine learning model for screening stable ABO_3 perovskites. Another key initiative is the Computational 2D Materials Database (C2DB) by Haastrup et al. [13], which includes results from large-scale density functional theory (DFT) calculations on approximately 2000 2D material variations, covering their structural, electronic, and magnetic properties. Similarly, the Crystallography Open Database (COD), curated by Gražulis et al. [14], hosts crystallographic data for over 380,000 organic and metal–organic compounds. The Open Quantum Materials Database (OQMD) by Saal et al. [15] includes crystal structure information for about 200,000 compounds obtained via high-throughput DFT simulations. These repositories have enabled the development of a wide array of machine learning (ML) and deep learning (DL) models for predictive materials science [16–18]. Complementing these repositories are Python-based open-source libraries such as MatMiner [19], which streamlines the querying of materials data from platforms like the Materials Project, Citrine, and the Materials Data Facility (MDF). The community has widely adopted MatMiner for data mining and the prediction of critical material properties [20–23]. Another powerful tool is Pymatgen [24], which facilitates the creation of DFT input files and provides seamless access to data from the Materials Project. Pymatgen has been instrumental in enabling AI-driven materials discovery, including the prediction and screening of materials properties [25–27]. Both MatMiner and Pymatgen offer comprehensive online documentation, including installation guides, tutorials, and example workflows, making them accessible to both novice and experienced users engaged in materials informatics.

10.3 Scalability of Machine Learning Models

Another challenge associated with data-driven approaches is their limited scalability when it comes to deploying frameworks for the generalized characterization of material systems. For instance, when the ML models are designed for a specific composition of alloy (AlCoCrFeNi) [5, 6], the extension of such ML models becomes difficult to characterize the wide variety of alloys consisting of other different elements. For better deployability of the ML/DL models, it is advisable to work with the generalized feature embeddings, such as Mat2Vec [28], composition-based feature vector (CBFV) [29], while developing the data-driven models. Although high-dimensional feature embeddings make the development of ML models computationally intensive, they enable greater model transferability and help address data diversity more effectively than using individual elemental spaces (the framework illustrated in Fig. 10.1). Such practices are particularly useful while characterizing the compositional materials such as alloys, perovskites, and bulk-metallic glasses. The computational cost associated with high-dimensional feature embeddings can be significantly reduced by leveraging pre-trained models—trained on these rich feature spaces—for transfer learning [8].

10.4 Materials Domain-Specific Awareness in Developed ML Models

The domain expertise in materials science is gained through years of experience and a deep understanding of the complex relationships between composition, structure, processing, and properties. Although general ML/DL models are often computationally efficient, they typically function as black boxes, making it challenging to interpret their predictions. One effective approach to addressing this limitation is to induce explainability of the developed ML model, which allows for the assessment of a model's behaviour by examining its dependence on specific input features. SHapley Additive exPlanations (SHAP)-based explanation of the ML model, which is a model-agnostic explainability technique, is one of the effective ways of capturing

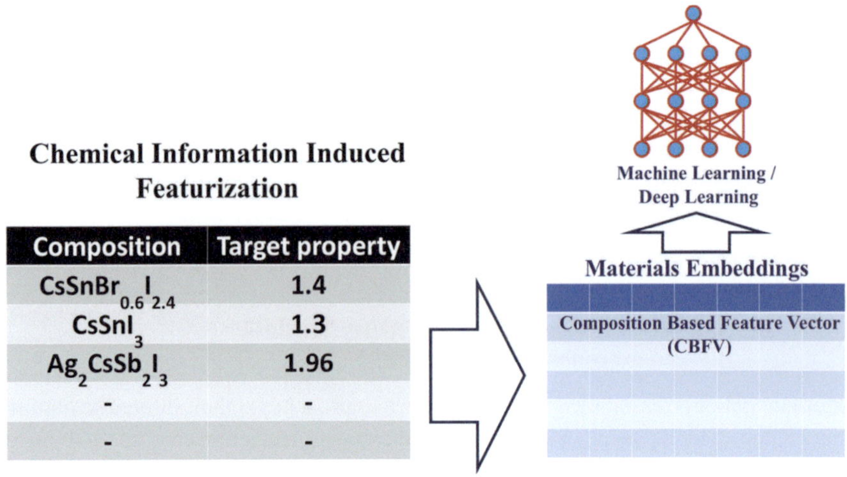

Fig. 10.1 Chemical domain-informed featurization for developing scalable ML/DL models

the underlying domain expertise. Gupta et al. [5] implemented SHAP-based explanations to interpret machine learning (ML) predictions for the characterization of the AlCoCrFeNi high-entropy alloy (HEA), aiming to design a lightweight alloy with enhanced stiffness. Traditionally reliant on expert intuition, alloy design in this study was guided by ML interpretability tools.

The SHAP analysis (Fig. 10.2) revealed that Al and Ni had the most positive influence on predicting Young's modulus and density, while Co had the least and often negative impact. Further analysis using SHAP force plots and interaction visualizations demonstrated that specific combinations of Al and Co concentrations significantly affected model performance. Overall, the study demonstrates how SHAP can provide valuable insights into element-wise contributions, enhancing the trust and interpretability of ML-driven materials design. Notably, the SHAP-based explanations effectively mirrored the decision-making process typically followed by human experts in conventional alloy development, highlighting its potential as a valuable tool for guiding rational alloy design.

10.5 Summary

This chapter outlines key challenges in building a materials informatics pipeline. One major hurdle—the lack of structured data—can be addressed through open-source data repositories, which are discussed here. Additionally, problems with scalability and generalizability in machine learning frameworks can be tackled by incorporating chemically informed features and leveraging transfer learning. Finally, to embed

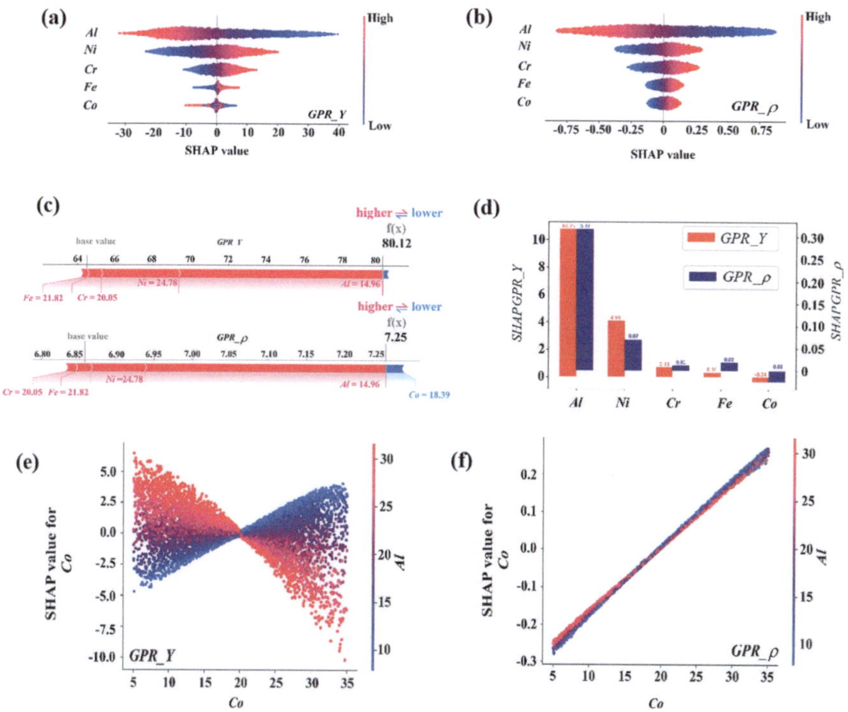

Fig. 10.2 SHAP-based explanation of ML models reported by Gupta et al. [5] to instil materials domain-aware computational decision-making for high-entropy alloy design

materials science expertise into these models, it's important to ensure and report their explainability.

10.6 Evolving Trends and Remarks

With the recent advent of Transformer architecture, the generative aspects of AI are being utilized in the materials science domain [30]. The fine-tuning of decoder-only model (a segment of Transformer) for generation of molecular structure [31], multi-agentic AI-driven molecular analysis [32], learning rare molecular events [33], generating high precision force fields [34], etc. are some of the recent examples of deploying generative AI in atomistic simulations. In addition to these efforts, text-mining is another domain, where GenAI is deployed for strategic knowledge extraction from the published scientific literature. In this regard, a few attempts have been made to develop materials domain knowledge-aware transformer architecture: for instance, Shetty et al. [35] developed MaterialsBERT by fine-tuning PubMed-BERT [36] with the large polymer corpora compiled by using a large number of abstracts of the published papers. Gupta et al. [37] fine-tuned SciBERT [38] with the corpus of the published literature on families of materials such as inorganic glasses, metallic glasses, and alloys and developed MatSciBERT. Wei et al. [39] proposed a transformer architecture (MatBERT) to construct, classify and property prediction for the compositional crystals. The probabilistic generation of crystals was validated with the DFT simulations. Huang et al. [40] developed MatInFormer, a generalized fine-tuned transformer architecture for materials property prediction. These few attempts focused on developing materials domain knowledge-aware transformer models, which required rigorous fine-tuning. These studies reported the successful deployment of GenAI in covering the different aspects of materials exploration, which highlights the promising potential of GenAI in materials science and engineering.

The development of hybrid models that integrate AI/ML with physics-based principles is anticipated to greatly enhance both the accuracy and clarity of material property predictions. These cutting-edge approaches are designed to deepen understanding of material behaviour [41, 42], support the design of materials with targeted properties, and close the gap between theoretical and experimental work—marking a significant step forward in atomic- and nanoscale materials research.

This book imparted the applied understanding of integrating machine learning (ML) frameworks with the conventional molecular dynamics (MD) simulation for nanoscale design of materials systems. The book particularly focuses on developing the predictive pipelines for large-scale characterization of materials and deployment of ML models for function-specific nanoscale design of the materials. The computational frameworks discussed in the book can be extended further for integrating generative artificial intelligence (GenAI) in materials design and discovery. In addition to that, the effective use of deep learning frameworks presents great potential in developing a scalable and versatile AI framework for materials design and discovery.

References

1. De Pablo JJ, Jones B, Kovacs CL, Ozolins V, Ramirez AP (2014) The materials genome initiative, the interplay of experiment, theory and computation. Curr Opin Solid State Mater Sci 18(2):99–117
2. Stach E, DeCost B, Kusne AG, Hattrick-Simpers J, Brown KA, Reyes KG, Schrier J, Billinge S, Buonassisi T, Foster I, Gomes CP (2021) Autonomous experimentation systems for materials development: a community perspective. Matter 4(9):2702–2726
3. de Pablo JJ, Jackson NE, Webb MA, Chen LQ, Moore JE, Morgan D, Jacobs R, Pollock T, Schlom DG, Toberer ES, Analytis J (2019) New frontiers for the materials genome initiative. NPJ Comput Mater 5(1):41
4. Gautam, Gupta KK, Bhowmik D, Dey S (2023) Probing the stochastic unconfined compressive strength of lime–RHA mix treated clayey soil. J Mater Civ Eng 35(3):04022469
5. Gupta KK, Barman S, Dey S, Mukhopadhyay T (2024) Explainable machine learning assisted molecular-level insights for enhanced specific stiffness exploiting the large compositional space of AlCoCrFeNi high entropy alloys. Mach Learn: Sci Technol 5(2):025082

6. Gupta KK, Barman S, Dey S, Naskar S, Mukhopadhyay T (2024) On exploiting nonparametric kernel-based probabilistic machine learning over the large compositional space of high entropy alloys for optimal nanoscale ballistics. Sci Rep 14(1):16795
7. Mohanty T, Chandran KS, Sparks TD (2023) Machine learning guided optimal composition selection of niobium alloys for high temperature applications. APL Mach Learn 1(3)
8. Jha D, Ward L, Paul A, Liao WK, Choudhary A, Wolverton C, Agrawal A (2018) Elemnet: Deep learning the chemistry of materials from only elemental composition. Sci Rep 8(1):17593
9. Jain A, Ong SP, Hautier G, Chen W, Richards WD, Dacek S, Cholia S, Gunter D, Skinner D, Ceder G, Persson KA (2013) Commentary: the materials project: a materials genome approach to accelerating materials innovation. APL Mater 1(1)
10. Ong SP, Cholia S, Jain A, Brafman M, Gunter D, Ceder G, Persson KA (2015) The materials application programming interface (API): a simple, flexible and efficient API for materials data based on REpresentational State Transfer (REST) principles. Comput Mater Sci 97:209–215
11. Dung NQ, Khoi LV, Khoa NL, Hai V, Thanh DX, Anh P, Oanh NTH, Tuyen ND, Duc NM, Thang LH, Long PH (2024) Unveiling phase changes in iron ore reduction: x-ray diffraction analysis with a user-friendly colab interface integrated with the materials project API. In: International conference on advanced materials and technology (ICAMT 2024), Hanoi Univeristy of Science and Technology
12. Liu H, Cheng J, Dong H, Feng J, Pang B, Tian Z, Ma S, Xia F, Zhang C, Dong L (2020) Screening stable and metastable ABO3 perovskites using machine learning and the materials project. Comput Mater Sci 177:109614
13. Haastrup S, Strange M, Pandey M, Deilmann T, Schmidt PS, Hinsche NF, Gjerding MN, Torelli D, Larsen PM, Riis-Jensen AC, Gath J (2018) The computational 2D materials database: high-throughput modeling and discovery of atomically thin crystals. 2D Mater 5(4):042002
14. Gražulis S, Merkys A, Vaitkus A (2020) Crystallography open database (COD). In: Handbook of materials modeling: methods: theory and modeling, pp 1863–1881
15. Saal JE, Kirklin S, Aykol M, Meredig B, Wolverton C (2013) Materials design and discovery with high-throughput density functional theory: the open quantum materials database (OQMD). JoM 65:1501–1509
16. Venkatraman V, Carvalho PA (2022) On the value of popular crystallographic databases for machine learning prediction of space groups. Acta Mater 240:118353
17. Veit-Acosta M, de Azevedo Junior WF (2021) The impact of crystallographic data for the development of machine learning models to predict protein-ligand binding affinity. Curr Med Chem 28(34):7006–7022
18. Souza A, Oliveira LB, Hollatz S, Feldman M, Olukotun K, Holton JM, Cohen AE, Nardi L (2019) DeepFreak: learning crystallography diffraction patterns with automated machine learning. arXiv preprint arXiv:1904.11834
19. Ward L, Dunn A, Faghaninia A, Zimmermann NE, Bajaj S, Wang Q, Montoya J, Chen J, Bystrom K, Dylla M, Chard K (2018) Matminer: an open source toolkit for materials data mining. Comput Mater Sci 152:60–69
20. Islam MM (2024) Low dimensional fragment-based descriptors for property predictions in inorganic materials with machine learning. arXiv preprint arXiv:2407.21146
21. Yazdani MH, Branicio PS, Nomura KI (2024) Benchmarking machine learning models on a dielectric constant database for bandgap prediction
22. Mazumder DD, Islam T, Ushna AN (2024) Band gap prediction of inorganic materials using artificial intelligence through feature transformation. Available at SSRN 4857921
23. Jiayuan D, Xiaoyu Y (2020) Integration and optimization of material data mining and machine learning tools. Front Data Comput 2(4):105–120
24. Ong SP, Richards WD, Jain A, Hautier G, Kocher M, Cholia S, Gunter D, Chevrier VL, Persson KA, Ceder G (2013) Python Materials Genomics (pymatgen): a robust, open-source python library for materials analysis. Comput Mater Sci 68:314–319
25. Peivaste I, Ramezani S, Alahyarizadeh G, Ghaderi R, Makradi A, Belouettar S (2024) Rapid and accurate predictions of perfect and defective material properties in atomistic simulation using the power of 3D CNN-based trained artificial neural networks. Sci Rep 14(1):36

26. Zhao H, Ezeh CI, Ren W, Li W, Pang CH, Zheng C, Gao X, Wu T (2019) Integration of machine learning approaches for accelerated discovery of transition-metal dichalcogenides as Hg0 sensing materials. Appl Energy 254:113651
27. Ong SP (2019) Accelerating materials science with high-throughput computations and machine learning. Comput Mater Sci 161:143–150
28. Tshitoyan V, Dagdelen J, Weston L, Dunn A, Rong Z, Kononova O, Persson KA, Ceder G, Jain A (2019) Unsupervised word embeddings capture latent knowledge from materials science literature. Nature 571(7763):95–98
29. Murdock RJ, Kauwe SK, Wang AYT, Sparks TD (2020) Is domain knowledge necessary for machine learning materials properties? Integrat Mater Manuf Innov 9:221–227
30. Jiang J, Ke L, Chen L, Dou B, Zhu Y, Liu J, Zhang B, Zhou T, Wei GW (2024) Transformer technology in molecular science. Wiley Interdiscip Rev: Comput Molec Sci 14(4):e1725
31. Bagal V, Aggarwal R, Vinod PK, Priyakumar UD (2021) MolGPT: molecular generation using a transformer-decoder model. J Chem Inf Model 62(9):2064–2076
32. Stewart I, Buehler MJ (2025) Molecular analysis and design using generative artificial intelligence via multi-agent modeling. Molec Syst Des Eng 10(4):314–337
33. Zeng W, Cao S, Huang X, Yao Y (2021) A note on learning rare events in molecular dynamics using lstm and transformer. arXiv preprint arXiv:2107.06573
34. Yuan S, Han X, Zhang J, Xie Z, Fan C, Xiao Y, Gao YQ, Yang YI (2024) Generating high-precision force fields for molecular dynamics simulations to study chemical reaction mechanisms using molecular configuration transformer. J Phys Chem A 128(21):4378–4390
35. Shetty P, Rajan AC, Kuenneth C, Gupta S, Panchumarti LP, Holm L, Zhang C, Ramprasad R (2023) A general-purpose material property data extraction pipeline from large polymer corpora using natural language processing. NPJ Comput Mater 9(1):52
36. Gu Y, Tinn R, Cheng H, Lucas M, Usuyama N, Liu X, Naumann T, Gao J, Poon H (2021) Domain-specific language model pretraining for biomedical natural language processing. ACM Trans Comput Healthcare (HEALTH) 3(1):1–23
37. Gupta T, Zaki M, Krishnan NA, Mausam (2022) MatSciBERT: a materials domain language model for text mining and information extraction. NPJ Comput Mater 8(1):102
38. Beltagy I, Lo K, Cohan A (2019) SciBERT: a pretrained language model for scientific text. arXiv preprint arXiv:1903.10676
39. Wei L, Li Q, Song Y, Stefanov S, Dong R, Fu N, Siriwardane EM, Chen F, Hu J (2024) Crystal composition transformer: self-learning neural language model for generative and tinkering design of materials. Adv Sci 11(36):2304305
40. Huang H, Magar R, Xu C, Farimani AB (2023) Materials informatics transformer: a language model for interpretable materials properties prediction. arXiv preprint arXiv:2308.16259
41. Zaeem MA, Thomas S, Kavousi S, Zhang N, Mukhopadhyay T, Mahata A (2024) Multiscale computational modeling techniques in study and design of 2D materials: recent advances, challenges, and opportunities. 2D Mater 11(4):042004
42. Sharma A, Mukhopadhyay T, Rangappa SM, Siengchin S, Kushvaha V (2022) Advances in computational intelligence of polymer composite materials: machine learning assisted modeling, analysis and design. Arch Comput Meth Eng 29(5):3341–3385

MIX
Papier aus verantwortungsvollen Quellen
Paper from responsible sources
FSC® C105338

If you have any concerns about our products,
you can contact us on
ProductSafety@springernature.com

In case Publisher is established outside the EU,
the EU authorized representative is:
Springer Nature Customer Service Center GmbH
Europaplatz 3, 69115 Heidelberg, Germany

Printed by Libri Plureos GmbH
in Hamburg, Germany